19

THE OPEN MEDIA PAMPHLET SERIES

THE OPEN MEDIA PAMPHLET SERIES

Weapons in Space

KARL GROSSMAN
Foreword by Dr. Michio Kaku

Open Media Pamplet Series Editor:
Greg Ruggiero

SEVEN STORIES PRESS / New York

A Seven Stories Press First Edition,
published in association with Open Media.

LIBRARY OF CONGRESS CATALOGING-IN-PUBLICATION DATA
Grossman, Karl.
  Weapons in space / Karl Grossman
  p.   cm.   — (Open media pamphlet series; 19)
  ISBN 1-58322-044-5 (paper)
  1. Space warfare. 2. Space weapons. 3. United States—Military policy.
I. Title. II. Series.
UG1530.G76 2000
355'.—dc21   00-032937

Book design by Cindy LaBreacht

College professors may order examination copies of Seven Stories Press
titles for a free six-month trial period. To order, visit www.sevensto-
ries.com/textbook, or fax on school letterhead to (212) 226-1411.

9 8 7 6 5 4 3 2 1

Printed in Canada.

Contents

Foreword
by Dr. Michio Kaku

When historians write the history of the 20th century, they will remark that the threat of all-out nuclear war, involving a cataclysmic exchange of tens of thousands of hydrogen bombs between the two superpowers, receded with the ending of the Cold War.

But just when one danger is fading, another one is rising ominously. Instead of ushering in an era of peace and prosperity, the beginning of the 21st century, historians will note, saw increased militarization, marked by the weaponization of outer space. They will remark that this represented a missed opportunity of enormous dimensions. Right before our eyes, the prospects of banning nuclear weapons is slipping through our fingers.

Unfortunately, most people are not aware of this. Vaguely hearing of arms control talks at the United Nations, people have been lulled to sleep, thinking that the great powers are finally dismantling their weapons.

Nothing could be farther from the truth. Sadly, the U.S. military is dangerously pursuing its goal of military superiority, even though there is not an enemy in sight.

The U.S. military is shadowboxing with itself.

The weaponization of space represents a real threat to the security of everyone on Earth. Not only will this squander hundreds of billions in taxpayer dollars, which are better spent on education, health, housing, and the welfare of the people, it will greatly accelerate a new arms race in space, with other nations working feverishly to penetrate a U.S. Star Wars program, or to build one themselves. A whole new round of the arms race could begin.

Ironically, it is the U.S. that stands to lose the most in such a race to militarize outer space. It is the U.S., not China or Russia, which is highly dependent on a vulnerable, fragile network of communication satellites. It is the U.S., not the developing countries, which has a high concentration of resources centered on just a handful of cities. In case of war, the U.S. would suffer greatly, its satellites blinded by anti-satellite weapons, its communications centers neutralized.

The time to stop this madness, therefore, is now, while Star Wars and affiliated programs are still in their infancy. That is why this book is so important. It raises people's awareness about a matter which is largely ignored by the established media. Once again, Karl Grossman has done a great public service in unmasking the true implications of weapons in space, which would not be shields of peace, but weapons of war. Mr. Grossman's efforts in alerting people to the true danger posed by the weaponization of space have greatly aided the cause for world peace.

Michio Kaku is Henry Semat Professor of Theoretical Physics at City University of New York

VISION FOR 2020: FULL SPECTRUM DOMINANCE

The United States is preparing to make space a new arena of war.

U.S. military documents speak of the U.S. seeking to "control space" and from space "dominate" the earth below. The U.S. military, furthermore, would like to base weapons in space. Billions of tax dollars are being poured annually into U.S. preparations for space warfare.

Is it a "return" of Star Wars? In fact, Star Wars, the popular name given to the Strategic Defense Initiative of President Ronald Reagan, never went away. With its enormously powerful complex of backers, it developed and maintained a momentum of its own.

With the assumption of power by George W. Bush and Richard Cheney and an administration intimately linked to corporate and right-wing interests committed to expanding space military activities, Star Wars has received a huge boost.

What the U.S. is up to is a violation of the intent of the Outer Space Treaty, the landmark 1967 international agreement that sets space aside for peaceful uses. Ironically, the U.S. was a leader in drafting this visionary treaty which seeks to keep war out of space.

There is only a narrow window to stop the U.S. plans from going forward and triggering what inevitably would follow: other nations meeting the U.S. in kind, an arms race, and ultimately war in space.

"If the U.S. is allowed to move the arms race into

space, there will be no return," says Bruce Gagnon, coordinator of the Global Network Against Weapons and Nuclear Power in Space, the organization that is internationally challenging U.S. preparations to turn space into a war zone, that has been striving at the grassroots to keep space for peace. "We have this one chance," he emphasizes, "this one moment in history, to stop the weaponization of space from happening."[1]

The U.S. space warfare plans are explicitly laid out in documents including the *Vision for 2020* report of the U.S. Space Command.[2] (The U.S. Space Command "coordinates the use of Army, Naval and Air Force Space Forces" and was set up by the Pentagon to "help institutionalize the use of space," notes its website http://www.spacecom.af.mil/usspace.)

The multicolored *Vision for 2020* features a laser weapon firing a beam from space zapping a target below. (Its cover is reprinted here because U.S. preparations for space warfare are so unbelievable, so incredible that it is best to see the actual documents firsthand. You can fully download this and many of the documents noted in this book from the U.S. Space Command website. The U.S. military is so brazen about its plans for space war, it displays them publicly on-line.)

Vision for 2020 starts with wording that crawls as in the beginning of the *Star Wars* movies: "US Space Command—dominating the space dimension of military operations to protect US interests and investment. Integrating Space Forces into war-fighting capabilities across the full spectrum of conflict."[3]

Vision for 2020, issued in 1996, compares the U.S.

GENERAL HOWELL M. ESTES III

"The increasing reliance of US military forces upon space power combined with the explosive proliferation of global space capabilities makes a space vision essential. As stewards for military space, we must be prepared to exploit the advantages of the space medium. This Vision serves as a bridge in the evolution of military space into the 21st century and is the standard by which United States Space Command and its Components will measure progress into the future."

US Space Command--dominating the space dimension of military operations to protect US interests and investment. Integrating Space Forces into warfighting capabilities across the full spectrum of conflict.

effort to control space and the earth below to how centuries ago "nations built navies to protect and enhance their commercial interests," how the great empires of Europe ruled the waves and thus the world.[4]

And *Vision for 2020* stresses the global economy. "The globalization of the world economy will also continue, with a widening gap between 'haves' and 'have-nots,'" says the U.S. Space Command.[5] The view is that by controlling space and the earth below, the U.S. will be able to keep those "have-nots" in line. The U.S. Space Command is readying itself to be "the enforcement arm for the global economy," says Bill Sulzman, director of Citizens for Peace in Space, the group challenging U.S. space military activities in Colorado Springs, Colorado, where the U.S. Space Command is headquartered.[6]

The U.S. military not only acknowledges—it proudly proclaims that U.S. corporate interests are involved in setting U.S. space military doctrine. President Dwight Eisenhower warned in his "farewell address" to the nation in 1961 of the "military-industrial complex." That linkage is stressed in the U.S. Space Command's *Long Range Plan.*[7]

The *Long Range Plan*, also starting out with a Star Wars-like type, states: "The *Long Range Plan* has been U.S. Space Command's #1 priority for the past 11 months, investing nearly 20 man-years to make it a reality. The development and production process, by design, involved hundreds of people including about 75 corporations."[8]

You don't need to do much investigating to find out the identities of those corporations. The *Long Range Plan* provides a list—beginning with Aerojet continuing through Boeing, Hughes Space, Lockheed Martin, Rand

Corp., Raytheon, Sparta Corp. and TRW to Vista Technologies.[9]

The *Long Range Plan*, issued in 1998, explains in its introduction that *Vision for 2020* "guides how our military space strategy will evolve in the 21st Century and is the standard for measuring the progress of USSPACE-COM [U.S. Space Command] and its components.... To carry out the Vision, we have developed a very ambitious and much needed *Long Range Plan*."[10]

"Now is the time," says the *Long Range Plan*, "to begin developing space capabilities, innovative concepts of operations for war-fighting, and organizations that can meet the challenges of the 21st Century.... Even as military forces have become more downsized in the 1990s, their commitments have steadily increased. As military operations become more lethal, space power enables our streamlined forces to minimize the loss of blood and national treasure.... Space power in the 21st Century looks similar to previous military revolutions, such as aircraft-carrier warfare and Blitzkrieg."[11]

"The time has come to address, among warfighters and national policy makers," the *Long Range Plan* goes on, "the emergence of space as a center of gravity for DOD [Department of Defense] and the nation. We must commit enough planning and resources to protect and enhance our access to, and use of, space. Although international treaties and legalities constrain some of the LRP's [*Long Range Plan's*] initiatives and concepts, our abilities in space will keep evolving as we address these legal, political, and international concerns."[12] Not to worry about international law, says the U.S. Space Command. It'll be taken care of.

The *Long Range Plan*—amid boxes containing quotes such as this from General Ronald Fogelman, Air Force chief of staff: "I think that space, in and of itself, is going to be very quickly recognized as a fourth dimension of warfare"[13]—makes a series of declarations:

> The United States will remain a global power and exert global leadership.... It is unlikely that the United States will face a global military peer competitor through 2020.... The United States won't always be able to forward base its forces.... Widespread communications will highlight disparities in resources and quality of life—contributing to unrest in developing countries.... The global economy will continue to become more interdependent. Economic alliances, as well as the growth and influence of multi-national corporations, will blur security agreements.... The gap between 'have' and 'have-not' nations will widen—creating regional unrest.... The United States will remain the only nation able to project power globally.... One of the long acknowledged and commonly understood advantages of space-based platforms is no restriction or country clearances to overfly a nation from space. We expect this advantage to endure.... Achieving space superiority during conflicts will be critical to the US success on the battlefield."[14]

And for those concerned about turning the heavens into a war zone, the *Long Range Plan* counsels: "Space has been 'militarized' for 40 years. Reconnaissance, surveillance, warning, communications, weather, and most recently, navigational satellites were designed and deployed to serve national security needs.... The increasing number of countries and consortia turning to space to provide and receive services—and to generate wealth—will force nations to adapt to this emerging environment."[15]

The *Long Range Plan* then continues on for more than 100 pages detailing U.S. plans for "Control of Space," "Full Spectrum Dominance," "Full Force Integration," and "Global Engagement."[16]

"Space is the ultimate 'high ground,'" says *Guardians of the High Frontier*,[17] a 1997 U.S. Air Force Space Command report. The Air Force Space Command is committed to "the control and exploitation of space," it says.[18]

"Master of Space" is a motto of the Air Force Space Command. "Master of Space" appears as a Space Command uniform patch displayed in *Guardians of the High Frontier* and is emblazoned on the front entrance of a major Space Command element, the 50th Space Wing in Colorado. Master of Space. That pretty well sums up the U.S. military attitude toward space.

Almanac 2000 is a just-issued Air Force Space Command report that flatly declares: "The future of the Air Force is space."[19]

"Into the 21st Century," it says, the U.S. Air Force needs to be:

> *Globally dominant*—Tomorrow's Air Force
> will likely dominate the air and space around

the world...*Selectively lethal*—The Air Force may fight intense, decisive wars with great precision, hitting hard while avoiding collateral damage in both 'real' space and in computer cyberspace. *Virtually present*—Space forces compliment [sic] the physical presence of terrestrial forces. Although they are not visible from the ground, space forces provide virtual presence through their ability to supply global mobility, control the high ground, support versatile combat capability, ensure information dominance and sustain deterrence. The future Air Force will be better able to monitor and shape world events..."[20]

U.S. military leaders are blunt in describing U.S. plans to make war in, from and into space. As General Joseph Ashy, then commander in chief of the U.S. Space Command, put it, "It's politically sensitive, but it's going to happen. Some people don't want to hear this, and it sure isn't in vogue, but—absolutely—we're going to fight *in* space. We're going to fight *from* space and we're going to fight *into* space," Ashy told *Aviation Week & Space Technology* in 1996. (Italics in *Aviation Week and Space Technology*.) "That's why the U.S. has development programs in directed energy and hit-to-kill mechanisms."[21]

In the article, headlined "USSC [U.S. Space Command] Prepares for Future Combat Missions in Space," Ashy spoke of "space control," the U.S. military's term for controlling space, and "space force application," its definition for dominating Earth from space. Said General Ashy: "We'll expand into these two missions because they will become increasingly important. We will engage terrestrial targets someday—ships, airplanes, land targets—from space. We will engage targets in space, from space."[22]

Or as then Assistant Secretary of the U.S. Air Force for Space Keith Hall, also director of the National Reconnaissance Office, told the National Space Club in 1997: "With regard to space dominance, we have it, we like it, and we're going to keep it."[23]

And as General Richard B. Myers, then commander in chief of the U.S. Space Command, in a speech titled "Implementing Our Vision of Space Control" delivered in 1999 to the U.S. Space Foundation, stated: "The threat, ladies and gentlemen, I believe is real. It's a threat to our economic well-being. This is why we must work together to find common ground between commercial imperatives and the president's tasking me for space control and protection."[24]

SPACE-BASED LASERS

Far more than reports and rhetoric are involved. There is a multibillion dollar project underway to build what was initially named the "Space-Based Laser Readiness Demonstrator," now called the "Space-Based Laser." The promotional poster for this laser shows it firing its ray in space while a U.S. flag somehow manages to wave in space above it.

The "Space-Based Laser" is considered by the military as a first step in space-based weaponry. It is a joint project of TRW, Boeing, Lockheed Martin, the U.S. Air Force and the Ballistic Missile Defense Organization. It "follows more than 15 years of TRW work developing technologies" for U.S. military-sponsored "space-based initiatives," declared a 1998 press release announcing the project. "It also complements work that TRW and Boeing have already done as members of Boeing-led Team ABL, which is developing the Air Force's first Airborne Laser system."[25]

In November 2000, the U.S. Department of Energy requested public comment on an "Environmental Assessment" for full development of the "Space-Based Laser." The development program "is estimated at $20-$30 billion," said the Public Affairs Office at the Army's Redstone Arsenal in Huntsville, Alabama in March 2000. It said Redstone, base of the U.S. Army Aviation & Missile Command, was among "four finalists" for the "Space-Based Laser test facility." A "team" including the

SPACE-BASED LASER READINESS DEMONSTRATOR

Preparing Today... To Protect Tomorrow

Aviation & Missile Command and the adjoining NASA's Marshall Space Flight Center was formed "to support the SBL program."[26] In December 2000, the Pentagon gave approval for the "Space-Based Laser" project to go ahead at NASA's Stennis Space Center in Mississippi.

Then there is a second space-based laser already in testing, the Alpha high-energy laser. Built by TRW, it conducted its twenty-second successful test firing on April 26, 2000. "In addition to producing about 25 percent more power than previous tests, Alpha generated an output beam that was almost perfectly round and more uniform in energy density," proclaimed after the firing a happy Dan Novoseller, TRW's Alpha Laser Optimization program manager.[27]

"Megawatt Laser Test Brings Space-Based Lasers One Step Closer," exclaimed *Space Daily*, the internet space website, about the test. It was "a significant step forward in the nation's disciplined maturation of the technology required to deploy the Space-Based Laser Integrated Flight Experiment." The article included a drawing of the Alpha laser with the caption: "Turning swords into lasers."[28]

Some six billion dollars-a-year—plus funds in the "black" or secret—are now going into U.S. space military activities. Much is being spent on what is now called U.S. Ballistic Missile Defense, what Reagan's Strategic Defense Initiative was renamed. Missile defense? In the fuller picture, what is sought is largely offensive.

Star Wars proponents regard missile defense—and have through the years—as a "layer" of a broad U.S. program for space warfare. The program is to be "multi-layered" and to include "theatre defense" (weaponry used in or in close proximity to an area of conflict) space-based weaponry and missile defense.

As Pulitzer Prize-winning author Frances FitzGerald concludes in her book, *Way Out There In The Blue: Reagan, Star Wars and the End of the Cold War*, published in 2000, Star Wars backers see an "initial deployment" of a missile defense system as "not an end in itself. In their view the 'thin' defense would have be thickened as time went on. 'It's better than having nothing,' Republican Representative Curt Weldon of Pennsylvania said of the Clinton program, but 'we're probably going to have to use space-based assets.' As always for the Republican right, the goal was weapons in space—that is, weapons which, if they materialized could contribute to an offense, as well as provide a defense for the United States."[29]

With the Bush-Cheney takeover, "the Republican right" with its "goal" of weapons in space, is now back in power.

PREVENTING AN ARMS RACE IN SPACE

Well aware of the U.S. space warfare plans, other nations of the world arranged for a vote in the United Nations General Assembly in New York on November 1, 1999, to reaffirm the Outer Space Treaty and, specifically, its provision that space be reserved for "peaceful purposes."

Some 160 nations voted for the resolution entitled "Prevention of An Arms Race In Outer Space." It recognized "the common interest of all mankind in the exploration and use of outer space for peaceful purposes" and reiterated that the use of space "shall be for peaceful purposes...carried out for the benefit and in the interest of all countries." The measure stated that the "prevention of an arms race in outer space would avert a grave danger for international peace and security."[30]

Only two nations refused to support the resolution: the U.S. and Israel. Both abstained. That stance was in line with a consistent U.S. pattern in international forums in recent times, of opposing efforts to keep space for peace as set forth in the Outer Space Treaty.

On November 20, 2000, the "Prevention of An Arms Race In Outer Space" resolution came again before the UN General Assembly and 163 nations voted in favor. Again the U.S. and Israel abstained, joined this time by Micronesia, a cluster of Pacific islands that depends on U.S. aid.[31]

Canada, certainly in no way a potential foe, has been highly active at the UN in seeking to strengthen the Outer Space Treaty with an agreement to ban all weapons in space. At a UN presentation in October 1999, Marc Vidricaire, counsellor of the Permanent Mission of Canada, noted that "Canada first formally proposed...a legally binding instrument" for a "ban of the weaponization of space" in January 1997 and "renewed our proposal" earlier in 1999. He cited in his speech the U.S. Space Command's *Long Range Plan* "including its recommendation to 'shape [the] international community to accept space-based weapons.'" The Canadian diplomat said, "Our objective is to ensure that pursuing the concepts of space control and force application are not extended by any state to include actual deployment of weapons in outer space."[32]

On October 19, 2000, Vidricaire was again at the UN sounding the alarm on behalf of Canada. "Outer space has not yet witnessed the introduction of space-based weapons. This could change if the international community does not first prevent this destabilizing development through the timely negotiation of measures banning the introduction of weapons into outer space," he said.[33]

"It has been suggested that our proposal is not relevant because the assessment on which it rests is either premature or alarmist," he said. "In our view, it is neither. One need only look at what is happening right now to realize that it is not premature...We have heard often before that there is no arms race in outer space. We agree. We would like to keep it that way for the sake of our own national security and for international peace and security as a whole." Vidricaire said: "There is also no question that no state can expect to maintain a monopoly on such knowledge—or such capabilities—for all time. If one state actively pursues the weaponization of space, we can be sure others will follow."

Russian President Vladimir Putin, in his first address at the UN, to the "Millennium Summit," on September 6, 2000, stated that "particularly alarming are the plans for the militarization of the outer space. In spring of 2001 we shall celebrate the 40th anniversary of the first flight of man to the outer space. That man was our compatriot, and we suggest to organize on that date, under the umbrella of the UN, an international conference on prevention of the outer space militarization."[34]

In December 2000, Canadian Prime Minister Jean Chretien and Putin, visiting Canada, issued a joint statement announcing that "Canada and the Russian Federation will continue close cooperation in preventing an arms race in outer space, including interaction in the preparation and holding in Moscow in the spring of 2001 of an international conference on the nonweaponization of outer space."[35] That conference, titled "Space Without Weapons," was held in Moscow between April 11 and 14, 2001.

UN Secretary General Kofi Annan declared at the opening of the UN Conference on Disarmament in Geneva, Switzerland in January 1999 that space must be maintained "as a weapons-free environment" and he urged "that we codify principles which can ensure that outer space remains weapons-free."[36]

Two months later at the UN in Geneva, a seminar was held on "The Prevention of an Arms Race in Outer Space" organized by the Women's International League for Peace and Freedom. It began with a presentation by me—an "Overview of the Current Stage of Militarization of Outer Space"—that included the information I've conveyed so far in this publication.

I concluded, "What the government of my country, the United States of America, is doing in space...gravely endangers life on this planet. It pushes us toward nuclear catastrophe. The military use of space being planned by the U.S. is in total contradiction of the principles of peaceful international cooperation that the U.S. likes to espouse. The aim is to develop a world in which it would literally be U.S.A. *uber alles.* This flies in the face of the spirit, the ideals of the United States of America. It denigrates those courageous men and women who came to this continent and fought the horrific evil of fascism in

World War II. It pushes us—all of us—toward war in the heavens."

I was followed by Wang Xiaoyu, first secretary of the Delegation of China, who declared: "Outer space is the common heritage of human beings. It should be used entirely for peaceful purposes and for the economic, scientific, and cultural development of all countries as well as the well-being of mankind. It must not be weaponized and become another arena of the arms race."[37]

The U.S., he said, "has over the years continued its efforts in developing space weapons with a view to deploying such advanced weapons in outer space in the near future. Huge amount[s] of human, material and financial resources have already been put into relevant plans and programs. The momentum has recently been greatly intensified. These ominous efforts will bring about the weaponization of outer space and lead to an arms race there."

The Chinese official cited the U.S. Space Command's *Long Range Plan*. "According to the plan, military space capabilities will become the major leverage in implementing national security and military strategies. Therefore, the priority task of the Space Force of the country [U.S.] in the 21st century is to gain and maintain space superiority," he said. "Its Space Command has thus put forward several operational concepts such as 'Control of Space' and 'Global Engagement.'"

"To put it simply," stated Wang Xiaoyu, the U.S. "is seeking to deploy in some years from now the ground-based interceptors which use outer space as a battlefield" as well as "weapon systems that are directly deployed in outer space, such as space operation vehicles, space-based platforms and lasers…. Thus people have come to realize that the weaponization of outer space has already become the sword of Damocles."

"Space domination is a hegemonic concept," he emphasized. "Its essence is monopolization of space and denial of others' access to it. It aims at using outer space for achieving strategic objectives on the ground." The result of the U.S. placing weapons in space, said Wang Xiaoyu, would be either that other countries would simply cave in and "acquiesce" to U.S. military dominance, or they would compete and further spread "weapons on the ground, sea, air and space."

He said that "against this background, the international community should act without any further delay to take effective measures, with a view to keeping the worst from happening." He said the UN "Conference on Disarmament, as the single multilateral disarmament negotiation forum, should live up to its obligations [and] negotiate and conclude legal instruments banning the test, deployment and use of any weapons, weapon systems and their components in outer space with a view to preventing the weaponization of outer space…. Let us work together to maintain a weapon-free and peaceful space for the 21st century."

The ambassador of Sri Lanka, H.M.G.S. Palihakkara, who is also special coordinator on the Prevention of an Arms Race in Outer Space for the UN Conference on Disarmament, then spoke declaring that "the international community should do something about this. It will be far too late even for the space powers to wait and let the technological momentum dictate the next stage of outer space weapons development. The moment you do research, there is the urge to deploy. The moment you deploy, others will deploy. Then we have a race on our hands."

Bill Sulzman, director of Citizens for Peace in Space, presented to the audience—which included many UN delegates—U.S. Space Command Major Kevin Kimble's 1998 lecture "to future Air Force officers" at the U.S. Air Force Academy. The lecture, appearing as an overhead on a screen, began: "There is a role for military use of space. Space is a medium useful for human endeavor. Human endeavor is accompanied by conflict. Human conflict, at its extreme, requires military solutions. Space is a medium requiring exploitation for military purposes. Space Control is the first order of business."

"It is clear from that presentation and from other recent Space Command printed materials that the U.S. does not recognize any international restrictions by treaty or otherwise to its military activities in space," said Sulzman. He cited *Vision for 2020*, The *Long Range Plan* and *Guardians of the High Frontier*. He spoke about a "most direct action of the U.S. in nullifying the 'peaceful purposes' concept of space law…its maintenance of a Space Warfare Center at Schriever Air Force base in Colorado Springs where future wars are being scripted with space as an area of conflict."

And Sulzman, a former Roman Catholic priest, ended by saying "I want to emphasize the positive. Many astronauts from many different countries have returned from their space travels to sing the praises of cooperation in keeping space free from human conflict. They point out the stunning beauty of the planet with its blue atmosphere and they always refer to the lack of borders and boundaries which separate us who live here below. Nationalism and militarism are the farthest things from their minds. We need to build on that spirit as we try to work together in the future to keep space as the common heritage of all humankind and reserve it for peaceful purposes."

Also making a presentation was Helen John of the Menwith Hill Women's Peace Camp who has been arrested many times for protests at the Menwith Hill military facility, a key command-and-control component of the U.S. space military program and a communications surveillance center, located in North Yorkshire in the United Kingdom.[38]

"We, who have protested outside the base at Menwith Hill at a Greenham-style women's peace camp for the past five years, cannot imagine how the international community can allow this to continue," said John. "This kind of power and this kind of spy technology is beyond Hitler's wildest dreams.... Following in the Fuhrer's footsteps, the U.S.A. now intends to dominate space to protect the American military and American business and commercial interests, having stolen everyone else's.... The U.S. Space Command documents *Vision for 2020* and The *Long Range Plan* spell out the far from peaceful future Uncle Sam has worked out for the rest of us. It involves increasing uses of deadly plutonium to produce the power sources for laser weapons.... Wernher Von Braun and Edward Teller's dream of orbiting battle stations is about to come true."

Women's International League for Peace and Freedom issued a statement declaring that "we see outer space as an integral part of the common heritage of humankind. All scientific exploration and any other use of outer space should be for civilian research only with a view to furthering the well-being of humanity and not for the destruction of life and the environment."

The following day, at the UN Conference on Disarmament, Li Changhe, ambassador for disarmament affairs of China, noting that the U.S. "in recent years has been intensifying its efforts in developing and testing weapons and weapons systems in outer space" thus making "prevention of an arms race in outer space [a] more pressing" issue, formally proposed "an international legal instrument banning the test, deployment and use of any weapons, weapon system and their components in outer space, with a view to preventing the weaponization of outer space."[39]

"My delegation requests that this be circulated as an official document of the Conference on Disarmament," said the Chinese diplomat, on the floor of the historic Council Chamber of the UN in Geneva, a room built for the UN's predecessor, the League of Nations, bedecked with murals illustrating the struggles and accomplishments of humanity. "We are ready to listen to the views and positions of all parties and make joint efforts in promoting substantive progress."[40]

From the floor, country after country declared their support.

However, the U.S. has refused to back this measure or the Canadian resolution to ban weapons in space. Because the Conference on Disarmament works on a consensus basis, U.S. obstinacy has frozen movement on this critical issue.

Ironically, it was the U.S. that was originally

involved in initiating the Outer Space Treaty, according to Craig Eisendrath, a former U.S. State Department officer who helped in its creation. Keeping space weapons-free was the original intent of the treaty, says Eisendrath. The Soviet Union had launched its Sputnik satellite in 1957 and "we sought to de-weaponize space before it got weaponized." A model the State Department used for its draft of the Outer Space Treaty, says Eisendrath, was the Antarctic Treaty which bars military deployments and states, "Antarctica shall continue forever to be exclusively used for peaceful purposes.[41]

The Soviet Union and the United Kingdom joined the U.S. in presenting the Outer Space Treaty, and it was adopted by the UN General Assembly in 1966. The Outer Space Treaty has now been ratified or signed by 123 nations. It entered into force in October 1967. The final wording of the treaty provided that "State Parties to the Treaty undertake not to place in orbit around the earth any objects carrying nuclear weapons or any other kinds of weapons of mass destruction, install such weapons on celestial bodies, or station such weapons in space in any other manner."

The intent of the Outer Space Treaty was "to keep war out of space," said Eisendrath, who became an educator after working for the State Department to and is now a senior Fellow at the Center for International Policy in Washington, D.C. He is a co-author of the forthcoming book, *The Phantom Defense: America's Pursuit of the Star Wars Illusion*. Eisendrath views the deployment in space of weapons such as the lasers that the U.S. military is delivering as "a violation" of the Outer Space Treaty.[42]

The resolutions put forth at the UN by Canada and

China, urged by Kofi Annan and supported by most nations of the world, would clear up any confusion and specifically bar any weapons in space. If approved, solid mechanisms could then be put in place to assure compliance, and violators would face international action. A huge problem could then be solved—in harmony with the foresight of those who developed the Outer Space Treaty. But that is not to the liking of the U.S. government today.

Mike Moore, then editor of *The Bulletin of the Atomic Scientists*, wrote in 1999: "The notion that the United States—or any country—might actually place weapons in space, as envisioned by the [U.S.] Space Command, is so repugnant that the United States ought to clearly repudiate it. Better yet, it should push to amend the Outer Space Treaty so as to definitively prohibit all weapons in space, not just weapons of mass destruction."[43]

But this is not the view shared by most of those in power in the U.S. in recent years and now. The same day that China introduced the resolution to ban all weapons in space at the UN Conference on Disarmament in 1999, I had a talk with a high U.S. official at the UN in Geneva. He was not happy with my remarks the day before at the seminar on "The Prevention of an Arms Race in Outer Space." On an anonymous basis, he sought to explain the U.S. stance.

In the wake of the Vietnam War, he said, the U.S. military believes it "can't field large numbers of ground troops." It was done for the Persian Gulf War, he noted as we stood in front of the U.S. Mission to the UN in Geneva—its entrance area strewn with razor-wire fencing and guarded by heavily-armed Swiss soldiers—but that took quite a build-up, "nine months of a drum roll." There is strong U.S. public resistance to having U.S. troops fight on the ground. But the U.S. military believes "we can project power from space," said the official, and that is why the U.S. military is moving in this direction.

There are profound moral, national and international issues raised by the determination of the U.S. to "control space" and from it "dominate" the earth, I noted, remarking that my uncles didn't go to Europe during World War II to fight for that kind of United States. Meanwhile, I said, I was not naive about the record of U.S. intervention in the affairs of other countries, pointing out that in the 1980s I wrote a book on U.S. efforts against Nicaragua (*Nicaragua: America's New Vietnam?*). But, I said, the U.S. seeking to control space "is way beyond the Monroe Doctrine."

Morever, if the U.S. moved ahead with this, would not other nations respond and an arms race in space follow? The official replied that the U.S. military has done analyses and determined that China is "thirty years behind" in competing with U.S. militarily in space and Russia "doesn't have the money."

I recounted to him my travels in China, seeing its technological power, and I pointed to its space prowess—how U.S. companies are now using China for launches. If the U.S. moves to "control space" and from it the earth, to deploy weapons in space, the Chinese would militarize space too—in something like three not thirty years, I suggested. And Russia might not have the rubles now, but it's a nation rich in natural resources and with enormous space abilities. A huge, potentially catastrophic miscalculation is being made, I said, heading the world toward war in the heavens. We parted in disagreement.

Later that day, I sat with a group of Chinese diplomats who indicated that if forced to do so, their nation could move into space militarily. "But we don't want to," said a young Chinese official. Rather, he said, China wants to use its resources to educate, house, feed and provide medical care for her billion people.

Echoing the myopic military view that the U.S. somehow can have an exclusive military use of space is George Friedman, a "defense expert" and co-author, with Meredith Friedman, of the 1997 book *The Future of War: Power, Technology & American World Dominance in the 21st Century*—yes, American World Dominance in the 21st Century. It's a detailed exposition on the proposition that the U.S. can dominate the earth for many years to come through the control of space.

"The age of the gun is over.... He who controls space controls the battlefield," said Friedman, in an interview, arguing that other nations "lack the money and/or technology to compete with us in the development of space-age weapons." He described China and Russia as "passing blips."[44]

His book, *The Future Of War*, concludes: "Just as by the year 1500 it was apparent that the European experience of power would be its domination of the global seas, it does not take much to see that the American experience of power will rest on the domination of space.... Just as Europe expanded war and its power to the global oceans, the United States is expanding war and its power into space.... Just as Europe shaped the world for half a millennium, so too the United States will shape the world for at least that length of time. For better or worse, America has seized hold of the future of war...."[45]

U.S. military plans for space will likely involve the use of nuclear power as an energy source for space-based weapons. The weapons the U.S. military is interested in deploying in space—especially lasers—will need large amounts of power and nuclear energy is seen as that power source.

As *New World Vistas: Air and Space Power for the 21st Century*, a 1996 U.S. Air Force Board report, states: "In the next two decades, new technologies will allow the fielding of space-based weapons of devastating effectiveness to be used to deliver energy and mass as force projection in tactical and strategic conflict.... These advances will enable lasers with reasonable mass and cost to effect very many kills.[46] This can be done rapidly, continuously, and with surgical precision, minimizing exposure of friendly forces. The technologies exist or can be developed in this time period."[47]

"Force application by kinetic kill weapons will enable pinpoint strikes on target anywhere in the world," says *New World Vistas*. "The Equivalent of the Desert Storm strategic air campaign against Iraqi infrastructure would be possible to complete in minutes to hours essentially on immediate notice."[48]

But there is a problem: space weapons would require large amounts of energy, and "power limitations impose restrictions" thus making "space-based weapons relatively unfeasible" now.[49] "A natural technology to enable high power is nuclear power in space.... Setting the emotional issues of nuclear power aside, this technology offers a viable alternative for large amounts of power in space," asserts *New World Vistas*.[50]

Solar power just cannot produce the necessary amount of energy needed for military purposes in space: "All solar collection systems in Earth orbit are limited by the solar constant of l.4 kilowatts per square meter," claims the report, and "large powers from solar collectors require large collection areas."[51]

Nuclear power "remains one of the attractive alternatives in generating large amounts of power in space," says the report,[52] declaring that "the Air Force should continue efforts toward making a safe nuclear reactor in space."[53]

The fifteen-volume report was prepared not only by U.S. military officers but, according to its appendix, high corporate, civilian and academic figures including, for its "Space Technology" volume, a Lockheed Martin vice president, NASA astronaut and manager from its Jet Propulsion Laboratory—Ronald Sega—and academics from MIT and Cornell.[54]

It was through investigating the use of nuclear power in space that I became aware of military plans to base weapons in space. In 1985 I learned that NASA intended to launch two space shuttles in 1986—one of the shuttles being the Challenger—with plutonium-fueled space probes aboard. After reaching orbit, the shuttles would launch the probes into space.

After reading about the plan in a Department of Energy publication, *Energy Insider*, I filed a Freedom of Information Act (FOIA) request with NASA, DOE and the national laboratories cited by *Energy Insider* as involved in the missions. *Energy Insider* said that the government had evaluated the consequences of an accident with the probes—on launch, in the atmosphere, or if a probe fell back to Earth—and I asked for this information.

It took nearly a year to get it. It was quite an uphill fight although FOIA requires that the government handle FOIA requests expeditiously. What the government finally advised was that, yes, there could be quite a disaster if the plutonium—considered the most toxic radioactive substance—was dispersed in an accident. But, DOE and NASA claimed, the chance was "very small...due to the high reliability inherent in the design of the Space Shuttle."[55] The likelihood of a catastrophic shuttle accident was put at 1-in-100,000.

On January 28, 1986, I was on my way to teach my Investigative Reporting class at the State University of New York, Old Westbury, when I heard over the car radio that the Challenger had blown up. I stopped at an appliance store and saw that horrible image on 100 TV sets—and thought, what if it was May of 1986, the date of the Challenger's next mission, when it was to have onboard the Ulysses plutonium-fueled space probe with 24.2 pounds of plutonium? There would have been many more lives lost if the explosion occurred then and plutonium was dispersed far and wide.

I began writing articles, then TV documentaries and a book (*The Wrong Stuff: The Space Program's Nuclear Threat To Our Planet*), on the use of nuclear materials on space devices. In the wake of the Challenger accident, NASA, incidentally, soon changed the odds of a catastrophic shuttle accident from 1-in-100,000 to 1-in-76. We only know real probabilities through empirical evidence.

Why use nuclear materials on space devices? For example, Ulysses was to be carried up by the Challenger and sent to orbit the sun. The plutonium on it and other space probes is used not for propulsion but to generate a small amount of electricity—256 watts on Ulysses— to power onboard instruments. Why not use solar energy? Why put the entire space program at risk by using nuclear material?

Part of the answer to that question was simple: as the informant "Deep Throat" told reporter Bob Woodward as he investigated the Watergate scandal—"follow the money." Who makes money on the use of nuclear devices in space? General Electric, which manufactured the plutonium systems, and, in recent years, Lockheed Martin, which took over that division of GE. Both GE and Lockheed Martin, it turns out, long lobbied the government to use their plutonium systems in space. Furthermore the national laboratories involved in developing space nuclear systems, such as Los Alamos National Laboratory and Oak Ridge National Laboratory, seek to retain and expand their funding.

Then I got to the military connection: the desire of the U.S. military to deploy nuclear-powered weapons in space. NASA was set up in 1958 as a civilian agency but, particularly after the end of the Apollo man-on-the-moon missions and its budget was cut, it became increasingly involved with the U.S. military. Indeed, the space shuttle program itself was created as a half-civilian, half-military program. The February 2000 mission of NASA's Endeavor space shuttle, for example, a flight to map the earth, was in large part a mission for the Pentagon, as some news reports mentioned but did not emphasize.

"By using two radar antennas—one on the end of the mast and a much larger one anchored in the cargo bay— scientists hope to obtain 3-D snapshots of Earth's terrain," reported the Associated Press in the last paragraph of its account. "And they expect those snapshots to be more plentiful and more accurate than any taken before" and they would be "a boon" to "the Pentagon."[56]

The U.S. military wants nuclear-powered weapons in space and that's been a key reason why NASA has been insisting on using nuclear power in space even when solar power would suffice. NASA coordinates its activities with the military.

Most recently, in 1997, NASA launched its Cassini space probe with more plutonium than ever used on a space probe—72.3 pounds. Afraid to use a shuttle for this launch, NASA sent Cassini up on a Titan-4 military rocket manufactured by Lockheed Martin. This Titan-4 made it up although three Titan-4's since have blown up on or soon after launching. Indeed, the Titan-4 launch record is now 1-in-10, one catastrophic accident for every 10 launches, a worse record than the space shuttle has.

Then, in what NASA admitted was the most dangerous phase of the Cassini mission, in 1999 it sent the Cassini space probe and its pounds of plutonium fuel back from space to buzz the earth. On August 17, 1999, NASA had Cassini whip by the earth at 42,300 miles per hour and 700 miles above it—in order to give it a "gravity assist" push to reach its final destination of Saturn. The good news: Cassini got past. It didn't dip down into the earth's 75-mile high atmosphere and break up, as NASA conceded it would have, for Cassini had no heat shield.

The bad news: on September 23, NASA's Mars Climate Orbiter seeking to pass over Mars came too close to the Martian atmosphere and crashed into the planet. That could have been Cassini crashing into the earth five weeks earlier. It turned out that the two teams of Mars Orbiter scientists—one a Lockheed Martin group, the other at NASA's Jet Propulsion Laboratory—were working with different scales of measurement: one feet, the other meters, and that's how the screw-up occurred. "Red-Faced Over The Red Planet: Metric Mixup Doomed Mars Spacecraft" was the headline of the Associated Press story, describing how "NASA's scientists' embarrassing failure to convert English units of measurement to metric ones...caused the navigation error.... 'It does not make us feel good that this happened,' said Tom Gavin of NASA's Jet Propulsion Laboratory."[57] Yes, accidents will happen when human beings are involved. Indeed, NASA does not consider human failure in its accident probability estimates because human stupidity can't be quantified.

More bad news: NASA, according to the U.S. General Accounting Office report, *Space Exploration: Power Sources for Deep Space Probes*, is "studying eight future space missions between 2000 and 2015 that will likely use nuclear-fueled electric generators."[58]

The next nuclear-fueled space mission is that of the Europa Orbiter, scheduled in 2003 to go to Europa, a moon of Jupiter (although NASA's Jet Propulsion Laboratory officials have been saying that the 2003 date for the Europa mission is "under review").[59]

The European Space Agency, ESA, meanwhile, has developed new "high efficiency solar cells" for use in space—as a substitute for nuclear power. And in 2003

ESA will be launching its Rosetta probe using solar arrays for power—to go beyond the orbit of Jupiter to rendezvous with a comet called Wirtanen. "Rosetta will make first contact with Wirtanen about 675 million km from the sun," notes ESA. That's 500 million miles from the Sun. "At this distance, sunlight is 20 times weaker than on Earth," ESA points out.[60]

But, again, NASA—seeking to coordinate its activities with the military and wanting to keep Lockheed Martin and the national nuclear laboratories in funds—sticks with nuclear power in space.

Accidents involving nuclear devices in space are not theoretical, they are real. It's not a sky-is-falling issue. Accidents have already occurred in the space nuclear programs of both the U.S. and the former Soviet Union, now Russia—in fact, there has been a 15% accident rate in both nations' space nuclear programs.

The most serious U.S. mishap occurred on April 21, 1964 when a U.S. navigational satellite (Transit 5BN-3) powered by a SNAP-9A (SNAP for Systems Nuclear Auxiliary Power) fueled with plutonium failed to achieve orbit and fell from the sky, disintegrating as it burned up in the atmosphere. The 2.1 pounds of plutonium scattered around the world. "A worldwide soil sampling program carried out in 1970 showed SNAP 9-A debris to be present at all continents and at all latitudes," according to the 1990 publication, *Emergency Preparedness for Nuclear-Powered Satellites*, a report by Europe's Organization for Economic Cooperation and the Swedish National Institute of Radiation Protection.[61]

Importantly, the type of plutonium used in space devices—Ulysses, Cassini, SNAP-9A and the others—is

Plutonium-238, which is 280 times "hotter" in radioactivity than the more widely known plutonium isotope, Plutonium-239, which is used in nuclear weapons.

Dr. John Gofman, professor emeritus of medical physics at the University of California at Berkeley, an M.D. and Ph.D. who developed some of the first methods of isolating plutonium for the World War II Manhattan Project, and co-discoverer of several radioisotopes including Uranium-233, has long connected the SNAP 9-A mishap to an increase of lung cancer on Earth. "Although it is impossible," he has said, "to estimate the number of lung cancers induced by the accident, there is no question that the dispersal of so much plutonium-238 would add to the number of lung cancers diagnosed over many subsequent decades."[62] The SNAP-9A accident caused NASA to become a pioneer in developing solar photovoltaic energy technology— solar panels that convert sunlight directly to electricity— now the power system on all U.S. satellites.

The worst Soviet space nuclear accident occurred on January 24, 1978 when Cosmos 954, a reconnaissance satellite powered by an onboard nuclear reactor, fell from orbit crashing into the Northwest Territories of Canada splattering nuclear debris over a huge area. "Eyewitnesses near the impact zone reported seeing a brilliant, glowing object accompanied by at least a dozen smaller growing fragments," according to *Emergency Preparedness for Nuclear Powered Satellites.* "During the first weeks of search, it became apparent that sizeable amounts of radioactive debris had survived reentry and was spread over a 600 km [kilometer] path from Great Slave Lake to Baker Lake."[63]

The most recent Russian space nuclear accident: the break-up of the Russian Mars 1996 space probe with a half-pound of plutonium aboard over the border region of Chile and Bolivia on November 16, 1996. [64]

Cassini carried the most plutonium of any space device so far.

And what a colossal disaster could have occurred if it had screwed up.

NASA in its *Final Environmental Impact Statement for the Cassini Mission* said that if the probe did not fly overhead as planned but dipped into the earth's atmosphere on the "flyby"—it would make an "inadvertent reentry," break up, and release plutonium and—these are NASA's words—"approximately 5 billion of the...world population at the time...could receive 99 percent or more of the radiation exposure."[65]

NASA, in its statement, said that if plutonium rained down on areas of natural vegetation, it might have to "relocate animals," if it fell on agricultural land, "ban future agricultural land uses" and, if it rained down on urban areas, to "demolish some or all structures" and "relocate affected population permanently."[66]

As to the human death toll: Dr. Gofman projected 950,000 dying as a direct result of a Cassini "flyby" accident. Dr. Ernest Sternglass, professor emeritus of radiological physics at the University of Pittsburgh School of Medicine, estimated the death toll at between 20 and 40 million people.[67]

The Outer Space Treaty has a provision that addresses damage caused by space devices. The treaty declares that a nation that launches "an object into outer space...is internationally liable for damage" caused by it. A follow-up 1972 UN treaty, "Convention on International Liability for Damage Caused by Space Objects," says "a launching state shall be absolutely liable" for such damage. But, in 1991, the NASA and the U.S. Department of Energy entered into a "Space Nuclear Power Agreement" to cover U.S. nuclear space flights with the Price-Anderson Act. This is a U.S. law which limits liability in the event of a nuclear accident to $8.9 billion for U.S. domestic damage and $100 million for damage to all foreign nations.[68]

Thus if an "inadvertent reentry" of Cassini back into the earth's atmosphere had occurred in 1999 and a part of Europe or Africa or Asia suffered nuclear contamination, all the nations and all the people affected could only have collected in damages—despite the land polluted, the number of people who would develop cancer—$100 million. The same will be true for an accident involving the plutonium-fueled Europa and other planned U.S. space nuclear shots ahead.

Meanwhile, in the event of an exchange involving nuclear-powered weapons in space, how would the resulting radioactive pollution affect life on Earth?

Furthermore, warfare in space will produce large amounts of space debris that could prevent humanity from journeying into space and limit us to the earth. The space above Earth is already littered with debris ranging from bolts and metal bits to defunct satellites and booster rockets. There are now 110,000 manmade objects orbiting in the space above the earth, some "8,870 larger than a softball," according to one recent survey.[69]

It has gotten so bad that NASA "now replaces pitted orbiter windows after most flights" of space shuttles, noted a 1997 report of the National Research Council, which warned that far more serious accidents involving space debris could occur resulting "in the loss of life or the vehicle."[70] It said: "The speed at which objects in low Earth orbit can collide makes these objects dangerous." The "typical impact velocities" of more than 20,000 miles per hour means that "even millimeter-sized objects can cause considerable damage."[71]

The U.S. Space Command monitors space debris and when the space shuttle is flown advises NASA. This Department of Defense "Space Surveillance Network (SNN) warns the space shuttle program of possible close conjunctions with catalogued orbiting objects. But probably more than 95 percent of the objects that could cause critical damage to the orbit are not catalogued because they are too small to be reliably detected by SSN detectors," said the report, Protecting the *Space Shuttle from Meteoroids and Orbital Debris*. It was done under a U.S. government contract involving the National Academy of Sciences and NASA.

The amount of "space junk" has doubled since 1990 and now poses "a navigational hazard" in space, says Norwegian space expert Erik Tandberg. The U.S. and Norway are planning a giant radar station, to be called Globus II, in Norway's Arctic, specifically to better monitor orbiting debris.[72]

An exchange involving space weaponry—a shooting war in space—would far surpass the amount of junk in orbit now above the earth. The heavens would be thoroughly littered.

EXCALIBUR

"My fellow Americans, tonight we are launching an effort which holds the promise of changing the course of human history," said President Reagan in a nationally-televised address on March 23, 1983, announcing the Strategic Defense Initiative later known as Star Wars.[73]

Several excellent books document Reagan's Star Wars plan, including, most recently, Frances FitzGerald's *Way Out There In The Blue: Reagan, Star Wars and the Cold War* and earlier works including *War In Space* by Nigel Flynn and two books by *New York Times* reporter William J. Broad.

They all tell of how, with the election of Reagan in 1980, Dr. Edward Teller, the principal figure in the formation and operation of Lawrence Livermore National Laboratory and the main figure, too, in developing the hydrogen bomb, saw a grand opportunity to push his plan for space warfare. The initial Teller program was to be centered around orbiting hydrogen bombs that would energize X-ray lasers. The concept was code-named Excalibur.

Flynn traces Reagan's exposure to Excalibur to a visit in 1967 to Teller's laboratory east of San Francisco when he was governor of California. Upon Reagan's election as president, Teller, an archconservative, worked with others in the U.S. right wing to push a program of space weaponization. Also involved was High Frontier, an organization set up, says Flynn, by the Heritage Foundation, a right-wing group in Washington. Teller was the 'sole scientist' with High Frontier," according to Broad's *Teller's War, The Top-Secret Story Behind The Star Wars Deception*.[74] The "idea of a space-based strategic policy began to take shape in the President's mind...assisted in 1981 by the formation" of High Frontier, says Flynn's *War In Space*.[75]

Teller insisted on Excalibur being a key element of the Star Wars scheme. As Broad explains the Excalibur concept in *Star Warriors*: "Around the H-bomb at its core are long, thin metal rods which, when struck by radiation, emit powerful bursts of X-rays. As the bomb at the core of an X-ray battle station exploded, multiple beams would flash out to strike multiple targets before the entire station consumed itself in a ball of nuclear fire."[76]

Former Army Lt. General Daniel O. Graham, ex-head of the Defense Intelligence Agency, advisor to Reagan and deputy director of High Frontier; however, balked at using hydrogen bomb-powered X-ray lasers as the base of Star Wars because "while other weapons could protect themselves, an X-ray laser waiting in orbit for an enemy attack would have to destroy itself in order to fire a beam at the attacker," and also because the U.S. "public would never accept the placement of nuclear weapons in space," according to Broad.[77]

Meanwhile, an array of other space-based weapons was being advanced, notably a variety of laser weapons as well as hypervelocity guns and particle beams. Most would have one thing in common: they'd require nuclear power, not actual hydrogen bombs like in the Teller Excalibur scheme, but nuclear power systems—"super" plutonium-fueled radioisotope thermoelectric generators and actual space reactors. There would also be additional "layers" including theatre defense and missile defense. Also part of the plan was a nuclear-propelled rocket using a "particle bed reactor," a design originated at Brookhaven National Laboratory on Long Island, New York. Code-named Timberwind, the nuclear-powered rocket would loft heavy Star Wars weaponry and other equipment into space. Later, it was given a second mission: serving as a rocket for flights to Mars.

The Pentagon would claim initially that "the emphasis is being given to non-nuclear weapons" in Star Wars, noted Flynn in *War In Space*. But, in fact, a massive amount of money was being spent on "nuclear-driven systems" for weapons. [78]

And by 1988, General James Abramson, director of the Strategic Defense Initiative Organization, was insisting that nuclear power was critical for Star Wars. He told a Symposium on Space Nuclear Power and Propulsion in Albuquerque, New Mexico in 1988—an annual event—that "without reactors in orbit [there is] going to be a long, long light cord that goes down to the surface of the earth."[79] "Failure to develop nuclear power in space," said Abramson, "could cripple efforts to deploy anti-missile sensors and weapons in orbit."[80]

Earlier, he began to let the nuclear space cat out of the bag by conceding at a Congressional hearing that "research on nuclear system technologies for both base load and potential weapons power" for Star Wars weaponry "is being conducted in conjunction with the Department of Energy." Abramson spoke of a "synergistic relationship" on Star Wars between the Pentagon and NASA. He noted that NASA, the Pentagon and the DOE were "cofunding" General Electric's SP-100 space reactor.[81]

POLITICS, PUPETEERING, AND "THE POWER STRUCTURE"

By 2000, the Star Wars plan with some changes—Teller's hydrogen bomb-Excalibur scheme was ultimately dropped as being wholly impractical—was still alive. "The heart of Ronald Reagan's 1983 Star Wars program lives on, kept beating by a mix of election-year politicking, behind-the-scenes defense-industry puppeteering and a fiercely committed group of conservative think tanks and antimissile-system advocates," *Time* magazine reported in a July 2000 article on missile defense.[82]

Those working to make sure Star Wars "kept beating" include the Republican right, aerospace corporations that have spent huge amounts of money in lobbying the political system, archconservative foundations like the Heritage Foundation and the U.S. military, especially its U.S. Space Command.

"Not surprisingly, noted *Time* in "The Reagan-era Star War Program Lives On," "Defense contractors...have a major interest in a NMD [National Missile Defense] system, especially since its ultimate cost is estimated at more than $30 billion. The four largest weapons contractors—Boeing, Lockheed Martin, Raytheon and TRW—together received more than $2.2 billion in missile-defense research-and-development money over a recent 21-month span, according to a report issued by the World Policy Institute. In 1997 and 1998, the latest years for which figures are available, Boeing, Lockheed Martin, Raytheon and TRW spent $35 million on lobbying."[83]

On the Republican right, no politician has been more active in recent times in keeping Star Wars alive than Senator Bob Smith of New Hampshire.

"Control of space is more than a new mission area —it is our moral legacy, our next Manifest Destiny, our chance to create security for centuries to come," Smith told the Aerospace Industry Association in 2000.[84]

"With the technology that we have already developed and demonstrated, we have the opportunity today to move forward," he stressed, on the program that "President Reagan envisioned almost 20 years ago, more than the marginal defense this administration has been struggling with for the past few months.... Space is absolutely critical to figure war fighting. This increasing importance was demonstrated in the Gulf War and the Balkans. I firmly believe that whoever controls space will win the next war." He advocated a broad program of including weapons in space and attainment by the U.S. of "global dominance."

Smith, a member of the Senate Armed Services Committee, has been the leader in Congress of creating a separate U.S. Space Force from the U.S. Air Force. "A Space Force would put the same muscle behind space missions that the Army, Navy and Air Force flex in their missions," he said. He authored an amendment to the National Defense Authorization Act for Fiscal Year 2000 that established a thirteen-member commission, chaired by Donald Rumsfeld, to look into how a U.S. Space Force would operate.

But so-called "centrist" political elements in the U.S.—including many elected Democrats—have been more than just going along. The massive amounts of money dispensed by the aerospace corporations has been one factor.

But there is more. Through the years a good number

of Democratic officials have backed a program of space warfare—even during the late 1980s when many Americans were refusing to accept the Reagan Star Wars scheme and through the Clinton administration of the 1990s.

"The Democratic Party has shown over and over again that it is in consonance with these plans for space warfare and global domination," says Bruce Gagnon of the Global Network Against Weapons and Nuclear Power In Space. "In Congress, they've supported—in strong numbers—the expenditures." Moreover, with NASA and U.S. corporate interests seeking to mine the moon, Mars and other planets along with asteroids, "you see a conjunction of interests."[85]

"It is not just right-wing kooks and the military promoting Star Wars," declares Gagnon. "It's what we can call the 'power structure.'"

Strong evidence of that can be found in the book *Military Space Forces: The Next 50 Years* that stresses on its title page that it was "Commissioned by Congress"— a Democratic-controlled U.S. Congress of the mid-1980s.

This blueprint for space warfare is as wild and extreme as anything produced by the U.S. Space Command or the Heritage Foundation, and yet was endorsed personally by a group of mostly Democrats and commissioned by a Democratic Congress. The list of officials signing off on the "Congressional Introduction" is topped by the facsimile signatures of Representatives Ike Skelton of Missouri and John Spratt of South Carolina— Democrats and leaders in recent years for missile defense. Then there are the signatures of then Senator John Glenn of Ohio, the ex-astronaut and a Democrat (given a NASA space shuttle ride in 1999); now U.S. Senator then Representative Bill Nelson, a Florida Democrat (representing Cape Canaveral and the rest of the "Space Coast" who got his NASA space shuttle ride in 1986); and Representative Harold Volkmer, a Missouri Democrat. The two Republicans are Representative John Kasich of Ohio and Ben Blaz, a non-voting member of the House from Guam.

The "Congressional Introduction" declares that Congress asked John M. Collins, senior specialist in national defense at the Congressional Research Service of the Library of Congress, "in June 1987 to prepare 'a frame of reference that could help Congress evaluate future, as well as present, military space policies, programs and budgets.'"

After a foreword by General John L. Piotrowski, then commander in chief of the U.S. Space Command, *Military Space Forces* opens with consideration of "economic and military enterprises" on the moon. "The moon is rich, in many natural resources.... iron, titanium, aluminum, manganese, and calcium are abundant.... Simple machines could easily strip top layers."[86]

Military bases on the moon would not only "defend" the mining operations but could take advantage of what *Military Space Forces* calls the "gravity well" of Earth. This is described as a channel in space between the moon and Earth. "Military space forces at the bottom of Earth's so-called gravity well are poorly positioned to accomplish offensive/defensive/deterrent missions, because great energy is needed to overcome gravity during launch," it says, but "forces at the top"—on the moon—could act "more rapidly. Put simply, it takes less energy to drop objects down a well than to cast them out. Forces at the top also enjoy more maneuvering room and greater reaction time."[87] A map of the best "site" on the moon from which the U.S. could take military advantage of this "gravity well" is provided and the work stresses that U.S. "armed forces might lie in wait at that location to hijack rival shipments"[88] of materials mined by other nations. The U.S., according to this Congressionally-authored plan, would engage in piracy in space.

Combat on the moon is discussed with the observation, "Lunar foxholes would provide better cover than terrestrial counterparts, because the absence of air confines blast effects to much smaller areas."[89]

Military Space Forces examines space weapons and states that nuclear weapons have a drawback. "Nuclear weapons detonated in atmosphere create shock waves, violent winds, and intense heat that can inflict severe damage and casualties well beyond the hypocenter." But in space "winds never blow in a vacuum, shock waves cannot develop...and neither fireballs nor superheated surrounding air develop above about 65 miles. Consequently, it would take direct hits or near misses to achieve required results with nuclear blast and thermal radiation."[90] On the other hand, "space is a nearly perfect laser environment...because light propagates unimpeded in a vacuum," it says.[91]

"Laser weapons, regardless of type (gas, chemical, excimer, free electron, solid state, X-ray), concentrate a tightly focused shaft or pulse of radiant energy photons on the target surface," *Military Space Forces* explains. "The beam burns through."[92]

The book also examines use of chemical and bio-

logical warfare in space and states: "Self-contained bios-
pheres in space accord a superlative environment for
chemical and biological warfare.... Clandestine opera-
tives could dispense lethal or incapacitating CW/BW
agents rapidly and uniformly through enemy facilities."[93]

"Conventional weapons" would have their place,
too, it says, pointing out that "high-speed
birdshot...could seriously damage most space facilities
which are strong enough to maintain structural integrity
and repel micrometerioids, but not much more."[94]

As to the UN Charter seeking "peaceful and
friendly" international relations, the Outer Space Treaty
designating space as a place where "exploration and
other endeavors 'shall be carried out for the benefit...of
all mankind,'" and the Moon Agreement of 1979 saying
"neither the surface nor the subsurface of the moon" or
"other celestial bodies within the solar system" shall
"become the property" of any person or state, *Military
Space Forces* declares: "The strength of such convictions
will be tested when economic competition quickens in
space."[95]

"Parties that hope to satisfy economic interests in
space must maintain ready access to resources on the
moon and beyond, despite opposition if necessary, and
perhaps deny access to competitors," it says.[96]

A good way to keep other nations from engaging in
space militarily, it goes on, is to "control attitudes" in
other countries. "Control over elitist and popular opin-
ion, using inexpensive psychological operations as a non-
lethal weapon system, could convince rivals that it
would be useless to start or continue military space pro-

grams," it says. "The basic objective would be to deprive opponents of freedom of action, while preserving it for oneself. Senior national executives, legislators, members of the mass media and, through them, the body politic, would be typical targets."[97]

Meanwhile, for the U.S., "Superiority in space could culminate in bloodless total victory, if lagging powers could neither cope nor catch up technologically."[98] As examples of the advantages of waging war from space, Collins states that "naval surface ships comprise" a particularly "inviting target category.... Former astronaut Michael Collins, who has been there and back twice, believes space is an ideal place from which to attack aircraft carriers and other major surface combatants."[99] And "strike forces on the moon could choose from the full range of offensive maneuvers."[100]

Military Space Forces also urges the use of nuclear power in space, both plutonium-fueled radioisotope thermoelectric generators and nuclear reactors which are "the only known long-lived, compact source able to supply military space forces with electric power about 10 kilowatts and multimegawatts.... Cores no bigger than basketballs are able to produce about 100 kw, enough for 'housekeeping' aboard space stations and at lunar outposts. Larger versions could meet multimegawatt needs of space-based lasers, neutral particle beams, mass drivers, and railguns."

Among the endorsements featured on the back cover of *Military Space Forces* are from then Senator Sam Nunn, a Georgia Democrat and chairman of the Senate Armed Services Committee, that, "This book will be an indispensable starting point," and then Representative

Les Aspin, a Wisconsin Democrat, later a secretary of defense under President Bill Clinton, stating: "No other military space study puts all pieces of the puzzle together."[101] General John W. Vessey, Jr., former chairman of the Joint Chiefs of Staff, states *Military Space Forces* "should be useful for decades."[102]

INTERNAL OPPOSITION

There are Democrats, of course, emphatically against the U.S. weaponization of space by the United States. One has been former U.S. Senator Charles Robb of Virginia, defeated for re-election in 2000, who declared in 1999: "The United States and other nations have rightly avoided placing weapons in space.... A space-based arms race would be essentially irreversible.... It defies reason to assume that nations would sit idle while the United States invests billions of dollars in weaponizing space, leaving them at an unprecedented disadvantage.... Once this genie is out of the bottle, there is no way to put it back in. We could never afford to bring all these systems back to Earth, and destroying them would be equally unfeasible, because the billions of pieces of space debris would jeopardize commercial satellites and manned missions." Moreover, said Robb, himself a former military officer, if space becomes a war zone "the fog of war would reach an entirely new density."[103]

And in the House of Representatives, leading opponents of Star Wars include Cynthia McKinney of Georgia, Lynn Woolsey of California and Dennis Kucinich of Ohio.

"Leave Star Wars to the movies," declared McKinney on the House floor April 11, 2000 as the three made an effort to stop the missile defense program. She spoke of the tens of billions of dollars that have been "been squandered on Star Wars. Now they have changed the name to National Missile Defense, but it is the same thing."[104]

"The U.S. Space Command calls for expanded war fighting capabilities in outer space. The guiding words in this country," said Kucinich, "ought to be 'thy will be done on Earth as it is in heaven,' not 'war be done in heaven as it is on Earth.' Let us work for peace on Earth, not war in space."[105]

Later that week, Kucinich gave the keynote address at the 2000 international meeting of the Global Network Against Weapons and Nuclear Power in Space. "We know that moving forward with a national missile defense system will set the stage for the advancement and proliferation of nuclear weapons in space," he declared. "And we know that once we continue down this road, we're going to be locked into funding an industry that makes missiles, and anti-missiles, and creates policies to promote the use of missiles, and more spending on missiles."[106]

He asked what happened to the "Cold War benefit. There's only a restless, ceaseless arms race which rides the newest technological wave, to continue to drain our national resources, to continue to create fear in America, to continue to create fear abroad, to continue to make the world less safe, and to continue to drive our national consciousness downward."[107]

Kucinich spoke of the opposition of the U.S. in the

UN to the resolution seeking to prevent an arms race in outer space by reaffirming the Outer Space Treaty. "It's my belief that the United States must sign on, and send a message to the world that space is for peace, not war.... Unfortunately, there are policy makers who are aimed at having the country make a statement totally the opposite."

He dismissed the claim by "advocates of the missile [defense] system that the system is not Star Wars all over again, it's ground wars.... Let's name names," he went on. "Lockheed Martin. TRW. Boeing. They now have a contract to build the space-based laser weapons that will be the follow-on technology to ballistic missile defense. This weapons system will enable the U.S. to have offensive capability in space, as called for by"— and he held up *Vision for 2020*.

"It seems that Orwell's vision of 1984 just follows the curve of time, and now it's taken us into 2020 where the vision of the United States Space Command is for war in space," said Kucinich. "Because what this states is that the U.S. Space Command intends to control and dominate space. Seize the heavens. The high ground."

"We are creating the seeds of the destruction of people all over the world, and it inevitably will come back to this country as well," he said to applause. "I think that we have a wonderful country. And I'm proud to be a member of the United States Congress serving out country," said Kucinich. "But if we love our country, then if we see our country taking a path that is dangerous, that we must help our country by challenging our country to do the right thing and not the wrong thing," he said to more applause.

"We have to address this as a moral issue as well," he said. "Because it is a moral issue. We're a country that should be about turning swords into plowshares.... Not in fashioning new, technologically superior swords of Damocles over the populations of the world. We cannot survive as a nation with that approach."

In the 1980s, too, there were members of Congress— not that many—who were vigorously challenging Star Wars. Some—such as Barney Frank, a Massachusetts Democrat—are still there. They introduced resolutions in 1983 and 1984 calling for international negotiations for "a ban on weapons of any kind in space." Neither bill even came close to passing.

And, in recent votes for missile defense, the numbers have been lopsided for deployment of missile defense, even after tests of the system have failed.

Now, with the Bush-Cheney takeover, the U.S. has an administration highly enthusiastic, gung-ho for Star Wars. Bush said during the presidential campaign that he would deploy missile defense "as soon as possible."[108] His administration would develop "a new architecture of American defense for decades to come."[109]

Wrote *Washington Post* reporter Walter Pincus: "His concept, which has encountered broad support among congressional Republicans, appears to be nearly identical to the missile shield proposed by his father."[110]

Bush has a "grandiose scheme for a real, all-out Star Wars scenario," as *Washington Post* columnist Mary McGrory described it.[111]

The Bush-Cheney team is loaded with links to the corporations most deeply involved in U.S. plans for waging war in space. Indeed, Cheney is a former member of the board of TRW.[112] His wife, Lynn Cheney, recently left the board of Lockheed Martin.[113]

"I wrote the Republican Party's foreign policy platform," Bruce Jackson, vice president of corporate strategy and development of Lockheed Martin, proudly told me in an interview. Jackson said he was selected to be "the overall chairman of the Foreign Policy Platform Committee" at the Republican National Convention, at which he was a delegate.[114] Thus the Bush administration will be using a foreign policy platform admittedly written by a top executive of Lockheed Martin, the world's biggest weapons manufacturer and central to U.S. space warfare preparations.

Jackson said that during the campaign he did not lead the advocacy for "full development of missile defense" because considering Lockheed Martin's heavy involvement, that "would be an implicit conflict of interest with my day job."[115]

Instead, he said, this was done by Stephen J. Hadley. Hadley, an assistant secetary for defense for international security policy in the administration of Bush's father, is a partner in the Washington law firm of Shea & Gardner—which represents Lockheed Martin.[116]

Hadley is a member, said Jackson, of "the Vulcans." This is the name given in the Bush campaign to an eight-member group which during the campaign advised Bush on foreign policy and also includes Condeleezza Rice, appointed National Security Council director.[117]

Jackson and Hadley have also worked closely on an entity called the Committee to Expand NATO. Jackson is president of this entity, based in the Washington office of the right-wing American Enterprise Institute; Hadley is its secretary.[118] Hadley was also a member of the National Security Council staff during the earlier Bush administration.[119] He has been appointed deputy director of the National Security Council in the new Bush administration.

"Space is going to be important. It has a great feature in the military," Hadley, speaking as "an advisor" to Bush, told the Air Force Association in an address at its national convention September 11, 2000. He stated that Bush's "concern has been that the [Clinton] administration's proposal does not do the job right and it doesn't reflect a real commitment to missile defense.... This is an administration that has delayed on that issue and is not moving as fast as he thinks we could."[120]

In the Bush choice of Donald Rumsfeld as U.S. secretary of defense, the U.S. got a man whom *The Washington Post* called the "leading proponent not only of national missile defenses, but also of U.S. efforts to take control of outer space."[121]

In 1998, a U.S. commission headed by Rumsfeld—most of its members appointed by then House Speaker Newt Gingrich and Senator Majority Leader Trent Lott, both archconservatives—reversed a 1995 finding by the nation's intelligence agencies that the country was not in imminent danger from ballistic missiles developed by new powers. The report by the Commission to Assess the Ballistic Missile Threat to the United States, which became known as the "Rumsfeld Commission," claimed that "rogue states" posed such a threat thus justifying prompt deployment of missile defense. Announcing his naming of Rumsfeld as defense secretary, Bush declared that he was "most impressed" by the "work" of the Rumsfeld commisssion. With Rumsfeld, a former defense secretary, back in that position again "we'll have a person who is...wise on the subject of missile defense." Bush went on: "We've got a great opportunity in America to redefine how wars are fought and won.... Our nation is positioned well to use technologies to redefine the military."[122]

Rumsfeld has been described by the avidly pro-Star Wars right-wing Center for Security Policy as a "trusted advisor" and a financial supporter and in 1998 was awarded its "Keeper of the Flame" award."[123] The Center's advisory board includes such Star Wars promoters as Edward Teller and Lockheed Martin executives, including Bruce Jackson.[124]

Says Gagnon of the Bush-Cheney administration: "This so-called election is a major victory for those who intend to put weapons into space at an enormous cost to the U.S. taxpayer and to world stability." He notes statements by Bush during the campaign that the U.S. should design and deploy "quantum leap weapons" and that Los Alamos and Sandia National Laboratories would play a major role in the development of weapons that will allow the U.S. "to redefine war on our terms."[125] Both laboratories have been deeply involved in space-based laser work.

Democrat Al Gore maintained during the campaign that any missile defense system should be limited in scope.[126]

A strong stand against missile defense and Star Wars was taken by Green Party candidate Ralph Nader and Socialist Party candidate David McReynolds.

The "power structure" of the U.S. gathers together at the Council for Foreign Relations, a 3,000 member group that includes among its members those who are considered the top figures in government—most of the now former leading officials in the Clinton administration, for example, Clinton's predecessor, George Bush, was a member along with many of his top aides.

Of the new Bush-Cheney administration, Council on Foreign Relations members include Cheney, Rice and Secretary of State Colin Powell. Stephen J. Hadley is also an extremely active council member. Indeed, in 2000 he gave speeches on missile defense at both the council's national conference and national program meeting. The presentation at the national program meeting was entitled, "Should the New Administration Deploy a National Missile Defense?"[127]

There are many members from U.S. banks and corporations—and media, too. Indeed, more than a dozen editors and writers of *The New York Times* alone are council members along with Diane Sawyer, Tom Brokaw, Dan Rather, Jim Lehrer, Barbara Walters, Katherine Graham, chairman of the *The Washington Post* and several of its top editors.[128] The elite council, founded in 1921, says its mission is "to serve our nation through study and debate, private and public."[129] In 1998, it issued a report titled *Space, Commerce, and National Security* written by Air Force Colonel Frank Klotz, described as a Military Fellow at the Council on Foreign Relations. "In summary," the report declared, "the most immediate task of the United States in the years ahead is to sustain and extend its leadership in the increasingly intertwined fields of military and commercial space. This requires a robust and continuous presence in space."[130]

HISTORICAL PRECEDENT

Another important element in the situation goes way back—to before Reagan unveiled Star Wars—and concerns how the U.S. space program began. "The space program of today has its roots deep in the strategy of world domination through global terror pursued by the Nazis in World War II," Jack Manno, a professor at the State University of New York, Environmental Sciences and Forestry College, pointed out in his seminal 1984 book *Arming the Heavens: The Hidden Military Agenda for Space, 1945-1995.* "Many of the early space-war schemes were dreamt up by scientists working for the German military, scientists who brought their rockets and their ideas to America after the war."[131]

Arming The Heavens details the development in Nazi Germany during World War II of the V-1 and V-2 rockets and how, at war's end, the U.S. sought to grab as many of the German rocket scientists as possible. "It was like a professional sports draft," Manno writes. And corporate America was deeply involved. Scientists from the Nazi Penenemuende Rocket Center "were turned over for inter-rogation to Richard Porter, who was in Germany repre-senting the General Electric Corporation, which held the Army contract for the first long-range ballistic missile under development in the United States." In the end, the U.S. "adopted nearly one thousand Germany military sci-entists, many of whom later rose to positions of power in the U.S. military, NASA, and the aerospace industry."[132]

"Wernher Von Braun and his V-2 colleagues...began working on rockets for the U.S. Army. They soon launched [at White Sands Proving Ground] in New Mexico the world's first two-stage rocket, using a salvaged V-2 as the first stage and a smaller booster rocket that fired when the first rocket burned out," Manno relates. "In 1949, with the beginning of the Korean War, the Army ordered Von Braun and his rocket team to the Redstone Army Arsenal at Huntsville, Alabama. They were given the task of producing an intermediate-range ballistic missile to carry battlefield atomic weapons up to two hundred miles. The Germans produced a modified V-2 renamed the Redstone."

Huntsville began to become a major center of U.S. space military activities—which it continues to be—and soon "Von Braun began to emerge as the most dynamic spokesman for America's budding space program."[133]

The U.S. military, on its Redstone Arsenal website, provides this narrative on Von Braun: "He became technical director of the Peenemuende Rocket Center in 1937, where the V-2 rocket was developed. Near the end of World War II, he led more than 100 of his rocket team members to surrender to the Allied Powers. Von Braun came to the United States in September 1945 under contract with the U.S. Army Ordnance Corps as part of Operation Paperclip. He worked on high-altitude firings of captured V-2 rockets at White Sands Proving Ground." Von Braun and his "group" were then sent to the Redstone Arsenal in 1949 where he became director of development operations. After the creation of NASA, "Von Braun and his team were transferred" to it "and became the nucleus of the George C. Marshall Space

Flight Center at Redstone Arsenal." For ten years Von Braun was Marshall's director, leaving in 1970 to go "to NASA Headquarters to serve as Deputy Associate Administrator."[134]

Former German Major General Walter Dornberger—who had been in charge of the entire Nazi rocket program—also becoming a powerful figure in the U.S. space program. "In 1947 as a consultant to the U.S. Air Force and adviser to the Department of Defense, Walter Dornberger wrote a planning paper for his new employees," relates Manno. "He projected a system of hundreds of nuclear-armed satellites all orbiting at different altitudes and angles, each capable of reentering the atmosphere on command from Earth to proceed to its target. The Air Force began early work on Dornberger's idea under the acronym NABS (Nuclear Armed Bombardment Satellites). As a variation on NABS, Dornberger also proposed an antiballistic-missile system in space in the form of hundreds of satellites, each armed with many small missiles. The missiles would be equipped with infrared homing devices and could be launched automatically from orbit. This concept was also taken under study by the Air Force in the 1950s. Labeled BAMBI (Ballistic Missile Boost Intercept), it was an idea that would reappear in the space-war dreams of the Reagan administration in 1983."[135]

Manno wrote in 1984: "The real tragedy of an arms race in space will not be so much the weapons that evolve—they can hardly be worse than what we already have—but that by extending and accelerating the arms race into the twenty-first century the chance will have been lost to move toward a secure and peaceful world.

Even if militarists succeed in arming the heavens and gaining superiority over potential enemies, by the 21st century the technology of terrorism—chemical, bacteriological, genetic, and psychological weapons and portable nuclear bombs—will prolong the anxiety of constant insecurity. Only by eliminating the sources of international tension through cooperation and common development can any kind of national security be achieved in the next century. Space, an intrinsically international environment, could provide the opportunity for the beginnings of such development."[136]

It is now the 21st century and Manno was saying from his home in Syracuse that in the past as today "control over the earth" is what those who want to weaponize space chiefly want.[137]

The Nazi scientists are an important "historical and technical link, and also an ideological link," he said. As to claims of space warfare being defensive—from how Reagan characterized his Star Wars plan as a "shield" to the appellation "missile defense" today, "it's all a smokescreen. The aim is to put all the pieces together and have the capacity to carry out global warfare including weapons systems that reside in space." A new element behind Star Wars is the development of a global economy and what is deemed a need by those promoting it to have "control over the process of globalization."[138]

GLOBALIZATION & CONTROL

The U.S. is now called a "unipolar superpower"—the only superpower left on Earth—and having supremacy over the world politically, economically and military is most important to the country's "power structure."

This "power structure"—more than the country's political far right—sees a U.S. that is overwhelmingly powerful militarily as required for globalization.

"The hidden hand of the market will never work without a hidden fist—McDonald's cannot flourish without McDonnell Douglas, the builder of the F-15. And the hidden fist that keeps the world safe for Silicon Valley's technologies is called the United States Army, Air Force, Navy and Marine Corps," wrote *New York Times* foreign affairs columnist Thomas L. Friedman in a March 1999 cover story in the magazine of what is considered the U.S.'s "paper of record." "What The World Needs Now: For Globalization To Work, America Can't Be Afraid To Act Like The Almighty Superpower That It Is," was the title of the piece. The full-cover illustration for it was a photograph of a clenched fist, with stars and stripes painted on it in red, white and blue.[139]

Large-scale ground war is now considered extremely difficult by the U.S.—just as the U.S. diplomat told me in Geneva. It's due to, as he said, what happened to the U.S. in Vietnam and it's also a result of TV's round-the-clock coverage when war breaks out these days giving people a picture of the face of war. Three GI's were captured during the war in the Balkans in 1998, for instance, and that became a marathon media event, provoking a national trauma before they were released. U.S. military strategy is thus now relying on "stand-off" weaponry: fighting wars from afar, with a seemingly bloodless, sanitized, video game-aura and a minimum of U.S. physical exposure, sending in stealth fighters and bombers and pushing buttons and, from hundreds, sometimes more than 1,000 miles away, firing off Cruise and Tomahawk missiles.

AGAINST PROLIFERATION AND ESCALATION

The use of space as a platform above an opponent's head from which the U.S. can project its power to Earth below is an extension of this new high-technology, long-distance way of waging war.

The central fallacy, however, the tragic miscalculation in the U.S. space warfare plans is that the U.S., even as a superpower, can—alone—"control" space and from it "dominate" the planet below. The U.S. can indeed move up into space with weapons. But how long will it be up there alone? In 1945, the U.S. also thought it had an exclusive—on the atomic bomb. That didn't last very long.

Once up in space with weapons, the U.S. will be joined by other nations: China,, Russia, India, who knows who else? No one will profit other than aerospace companies and weapons manufacturers—Boeing, Lockheed Martin, Raytheon, TRW and the rest. The U.S. will have brought armed conflict up into space. What a legacy for our children and their children! It will be *Star Wars* but for real. And, it would be violent. It would not create less bloodshed than a ground war. It would not play well on TV. It would end up being expensive beyond all belief. It could pollute space with enough debris to make make future space missions impossible. And it would pass on a lethal radioactive legacy here on Earth for generations to come.

"One of the most important decisions affecting the next 1000 years will be whether to weaponize space or to advance the peaceful uses of space," wrote Patricia Mische, now the Lloyd Professor of Peace Studies and World Law at Antioch College in her important book *Star Wars and the State of Our Souls: Deciding the Future of Planet Earth.*[140]

"Powerful economic, political, technocratic, military and other forces want to weaponize space," she wrote back in 1984—and her words and the situation are as pressing today. "To do so would virtually close the gateway to the peaceful uses of space. These forces are applying old mentalities of war-fighting, national self-sufficiency, and dominance that may have been acceptable in bygone ages, but endanger the earth and human survival in the new area of nuclear weapons and global economic, communications, and environmental interdependency."

"We stand at a critical moment of history when the decision we are about to make with regard to outer space will affect everything that is to follow for thousands of years," said Mische. "A failure to find the right response now will be a failure of such immense magnitude that we and the planet may never recover from our mistake."

The Global Network Against Weapons and Nuclear Power was founded in July 1992 with the inaugural meeting of 200 held in Washington D.C.'s City Council Chambers.

There was Suzanne Marinelli of the Sierra Club of Hawaii who was fighting the establishment on her island of Kauai—"it really is paradise"—of a Star Wars launch site. (The site was established and has been the scene of numerous protests.) Chris Brown of Citizens Alert in Nevada, where plans were being made to test the "Timberwind" nuclear-propelled rocket, was there and spoke of the toll of atomic testing on the people of Nevada, Utah and Montana. It is time "to stop making the citizens" of the western U.S. "victims of the country's addiction to nuclear weapons and nuclear power," said Brown. (The "Timberwind" nuclear-propelled rocket was cancelled two years later but in recent years, the Pentagon and NASA have resumed development of a nuclear-propelled rocket with much of the work going on at the University of Florida in Gainesville, Los Alamos National Laboratory and NASA's Marshall Space Flight Center in Huntsville.) Canadian scientist Walter Dorn was there complaining about the U.S. opposing UN guidelines on space activities involving nuclear power.

"We're going to educate, mobilize, activate and influence the whole space policy program and turn the U.S. away from weaponizing and nuclearizing space," said Gagnon at a press conference at which an "action" platform was outlined. It called on the U.S. Congress to "zero out space-based weapons fund" and the budget for "Timberwind." It declared support of Iowa Democratic Senator Tom Harkin's Outer Space Protection Act banning weapons in space (which, like the measures in the House in the 1980s, failed). The platform said the Global Network would work "to remove the secrecy and expose the misinformation presented to the public regarding these issues" and concluded by saying that the Global Network would seek "to build international cooperation in space and urge the reallocation of resources to benefit all humanity and to protect the planetary environment."[141]

Since 1992, Gagnon relates, the Global Network "has met each year in order to bring together activists who are working on, or are interested in, space. It was the intention of the founders to create an organization that would serve as a clearinghouse for issues surrounding the nuclearization and weaponization of space, and act as a spark to ignite education and organizing in order to build an international citizens movement."

Annual meetings have been held in locations including New Mexico, Florida, England, Germany and Colorado—indeed, in 1993, the organization combined its annual meeting with a protest as Edward Teller spoke before the National Space Foundation in Colorado Springs and also picketed the U.S. Space Command headquarters there. "It is a tradition that each Global Network meeting include a public demonstration at a space-related facility or event," says Gagnon.

The 2000 meeting took the form of four days of "protest events" beginning with demonstrations at the U.S. Treasury Department—stressing how much money is being spent by the U.S. on space military activities— and right next to the Treasury Department, in front of The White House. The next day, "Star Wars Revisited: An International Conference on Preventing an Arms Race in Space" began at American University, keynoted by Kucinich. Among the speakers were: Dr. Helen Caldicott, president emeritus of Physicians for Social Responsibility; Professor Patricia Mische; Bill Sulzman and Loring Wirbel of Citizens for Peace in Space; William Hartung of the World Policy Institute; Carol Rosin of the

Institute for Security and Cooperation in Outer Space; Helen John; Regina Hagen of the Darmstaedter Friedensforum in Germany; Alice Slater, president of Global Recource Action Center for the Environment; David McReynolds of the War Resisters League; and South African poet and former co-prisoner with Nelson Mandela, Dennis Brutus.

The next day, Global Network members joined activists challenging the World Trade Organization and the global economy in their April 16, 2000 demonstration in Washington. The Global Network members pressed the relationship between globalization and the U.S. space warfare plans. And, the following day, Global Network members went to offices at the U.S. Congress.

The 2001 international meeting of the Global Network was held in Leeds, England in May and the 2001 U.S. national meeting in March in Huntsville, Alabama. On October 13, 2001, the Network will be holding an "International Day of Protest to Stop the Militarization of Space."

"From the outset we have tried to identify and activate people living near the key Space Command and/or NASA facilities around the world as the first steps in building a global grassroots movement to keep space for peace," said Gagnon. Thus early leaders in the movement, he noted, came from Colorado Springs (U.S. Space Command Headquarters); Leeds in England (Menwith Hill and also the nearby Fylingdales U.S. bases); and Darmstadt, Germany (the operational headquarters of the European Space Agency); and Florida (Cape Canaveral). Gagnon, a former organizer for the United Farmworkers Union and a U.S. Air Force Vietnam-era

veteran, was for fourteen years coordinator of the Florida
Coalition for Peace and Justice. The Global Network in
its early years was based out of the office of the Florida
Coalition.

The Global Network, further, he said, has grown to
include "activists from communities near other important space installations" including Vandenberg Air
Force Base in California (the launch site for National
Missile Defense tests); Kirtland Air Force Base in Albuquerque, New Mexico; Fort Meade, Maryland where the
National Security Agency is based; Marshall Space Flight
Center in Huntsville "and a host of corporate aerospace
facilities like Lockheed Martin" (Valley Forge, Pennsylvania and other locations) "and Raytheon" (Tucson, Ari-
zona).

"Among the primary goals of the Global Network,"
said Gagnon, "has been to educate and create enough
momentum that other peace groups, long working hard
on issues of nuclear disarmament, would understand
that without a strong coordinated movement to stop
plans for space weaponization, there could never be any
real opportunity to abolish nuclear weapons. By presenting workshops at conferences of most national peace
groups, the Global Network has intensified the con-
sciousness-building process about space weaponization
with the movement's most natural allies."

"Looking at the voting records of Congress over
recent years reveals that an organizing strategy based in
Washington D.C. is not one that will bear significant
fruit in the foreseeable future," he said. "Even with the
unwavering support of congresspersons like Representative Kucinich and others, the Global Network realized

that it is the grassroots that must ultimately be moved to prevent an arms race in space. Thus the Network has committed its time and resources to educating and mobilizing the public, knowing that in time, if the people can be empowered to lead, Congress and the White House will follow."

"Similarly," he said, "at the outset the Global Network came to understand that the U.S. peace movement alone could never lead this movement to keep space for peace. Because the issue of warfare in space has universal implications, the Global Network has taken the issue to every part of the world."

"It is with great pride," he said, "that the Global Network points to active support in places like Bangladesh, England, Ghana, Germany, Romania, Australia, China, India, Japan, Egypt, Azerbijan, Nepal and South Korea. This growing movement indicates that people all over the world understand the U.S. is now poised to plant the bad seed of war, greed, and environmental degradation in space. Citizens in these places rightly see themselves as part of the solution to this problem and have joined the struggle to maintain and strengthen the United Nations' treaties—the Outer Space Treaty of 1967 and Moon Agreement [which, among other things, forbids military bases on the moon]. "The idea that space must become a protected environment has caught on worldwide!"

"Creating a global democratic debate about the kind of seed that humankind should carry into space is the ultimate goal of the Global Network," said Gagnon. "The vision of people worldwide gazing at the moon and stars, sharing this tiny planet in space, and speaking with a collective voice that calls for protecting space from the evils that we have sown on this earth, is the true work of the Global Network. Members of the organization do not accept that we must continue to squander hundreds of billions of dollars on research and development for space-based lasers, anti-satellite weapons and nuclear-powered rockets. The time has come to say enough is enough!"

Gagnon offers the following "suggestions of what you can do to join this historic movement for peace in space.

• Learn more about the U.S. scheme to "control space" and from it "dominate" the earth. An "organizers packet" of information can be ordered from the Global Network for $5.

• Educate others by holding house meetings, writing letters to the editor, organizing teach-ins, and taking friends to talk with your political representatives.

• Organize public events and protests that call for peace in space. Invite speakers from the Global Network to attend.

• Check the Global Network's website regularly and get on its e-mail subscription list. Check out the large web-links section (including U.S. Space Command sites) on the Global Network website: <www.space4peace.org>

• Become an individual member or group affiliate of the Global Network.

• Send representatives from your community to upcoming meetings and protest events being organized by the Global Network. Check its website calendar for details.

The address of the Global Network Against

Weapons and Nuclear Power in Space is P.O. Box 90083, Gainesville, Fl. 32607. Its phone number is (352) 337-9274. Its E-mail is: <globalnet@mindspring.com>

The Global Network's board includes representatives from around the world including Ghana, United Kingdom, Romania and Germany. Board member Regina Hagen from Darmstadt, Germany recounts that "when I first heard about the nuclearization and the weaponization of space, I was shocked, I couldn't believe it. I knew about the old Strategic Defense Initiative back in the '80s but then I thought that passed because the Soviet Union was no longer there and no longer was a threat to the U.S." But, said Hagen, a scientific translator, the U.S. military "just keeps going on with their plans and with planning for war in space and war from space and war to space."[142]

She recalled that while attending a Global Network meeting in Colorado Springs, she went to a briefing given by an officer from the U.S. Space Command. "One of the very early slides" shown by the officer "said 'control of space," noted Hagen. "I asked…how would you as an American feel if we Europeans, or even we Germans, would claim that space is ours, that we want to control space, that we want to dominate space, that the resources in space are ours—I mean how would you feel? This is so unbelievable. Why should the Americans have the right to control space which is meant to be there for everyone?"

"The United States is trying to find a new bully mission in the aftermath of the Cold War," says Loring Wirbel of Citizens for Peace in Space. "We already have the largest economy on the planet, we want to be able to control resources, to control the balance between have and have-nots, to control who stays wealthy and who remains poor. And that implies being able to constantly monitor the planet, constantly challenge anyone who would even dare to question our dominance of the planet, and that ultimately means that the arms control treaties that were in place many years ago no longer apply because we intend to be the unipolar superpower."[143]

The U.S., says Wirbel, "needs to express its leadership through good works and good examples. The more we try to achieve dominance through wielding power and having our own way all the time, the more we lose the essence of our democracy that makes us an exceptional nation and the more we move toward this dominance regime, the more I have to say I'm embarrassed to be an American."

In January 2001, the report of the second U.S. commission headed by Donald Rumsfeld—the Commission to Assess United States National Security Space Management and Organization—was issued. Calling for "power projection in, from and through space," chaired by the incoming defense secretary, it broadcast the full commitment of the new U.S. administration to space warfare.[144]

"We know from history that every medium—air, land and sea—has seen conflict," declares the report. "Reality indicates that space will be no different. Given this virtual certainty, the U.S. must develop the means both to deter and to defend against hostile acts in and from space. This will require superior space capabilities."[145]

"In the coming period," states the report, "the U.S. will conduct operations to, from, in and through space in support of its national interests both on the earth and in space."[146]

And the commission urges that the U.S. president "have the option to deploy weapons in space to deter threats to and, if necessary, defend against attacks on U.S. interests."[147]

The 13 members of this Rumsfeld commission included two former commanders in chief of the U.S. Space Command and an ex-commander of the Air Force Space Command along with retired Republican U.S. Senator Malcolm Wallop. The report's thumbnail biography of Wallop notes he "is currently a Senior Fellow with the Heritage Foundation" and "in 1977 he was the first elected official to proposed a space-based missile defense system."[148]

The report continues the spin of using "missile defense" to sell what is, in fact, a broad program of space warfare. It warns of a "Space Pearl Harbor." A reading of this report, like *Vision for 2020, Long Range Plan* and the other space military plans, reveals the sweeping extent of what is involved. It is "possible to project power through and from space in response to events anywhere in the world," it stresses. "Unlike weapons from aircraft, land forces or ships, space missions initiated from Earth or space could be carried out with little transit, information or weather delay. Having this capability would give the U.S. a much stronger deterrent and, in a conflict, an extraordinary military advantage."[149]

Senator Bob Smith, whose legislation established the commission, wanted it to consider his scheme for a separate U.S. Space Force and the commission recommends a "Space Corps" like the Marine Corps[150] and a possible "transition" to a fully separate Space Force or "Space Department"—on par with the Army, Navy and Air Force—in several years hence.[151]

The report has a section on the U.S. getting around laws such as the Outer Space Treaty covering space military activities. It emphasizes: "There is no blanket prohibition in international law on placing or using weapons in space."[152]

The new report was immediately applauded by the U.S. military. "The Air Force welcomes the Space Commission's report and is enthusiastic about the observations and recommendations that determined a realigned and rechartered Air Force is best suited to organize, train and equip space forces," said a dispatch on the U.S. Air Force website. It quoted Brigadier General Michael A. Hamel, space operations and integration director, as saying: "This is a golden opportunity for the Air Force to create a strong center of advocacy and commitment to national security space efforts."[153]

"The Air Force," Reuters reported, "embraced a report that cited the 'virtual certainty' of future hostile action in space and said it was moving forward with plans to boost U.S. military strength in the heavens." Major General Brian Arnold, director of space and nuclear deterrence, was quoted as saying: "The Air Force strongly supports the Space Commission report and is already moving to implement many of its recommendations."[154]

With the Rumsfeld report in hand, the Bush administration is moving fast and hard. "In a matter of weeks and without presidential appointment or decree, the nation's policy toward space appears to have shifted from one of civilian exploration and commercial exploration to one dominated by war fighters," noted *Florida Today* in February 2001. "In the weeks since President Bush was sworn in, four-star generals and their aides have stepped forward to flex their new-found muscle, driven largely by recommendations contained in [the] recent report."[155]

Kofi Annan, in opening the Third United Nations Conference on Exploration and Peaceful Uses of Outer Space, declared: "Above all, we must guard against the misuse of outer space. We recognized early on that a legal regime was needed to prevent it from being another arena of military confrontation. The international community has acted jointly, through the United Nations, to ensure that outer space will be developed peacefully."[156]

"But there is much more to be done. We must not allow this century, so plagued with war and suffering, to pass on its legacy, when the technology at our disposal will be even more awesome," said Annan. "We cannot view the expanse of space as another battleground for our earthly conflicts."

But, as the new century begins, that is not the vision of the "power structure" of the United States of America.

There must be strong opposition in the U.S., international action, and a global agreement banning weapons in space. *We must keep space for peace.*

"The people of the U.S., the people of the world, must learn," says Gagnon, "about what the U.S. is up to—and stop it."[157]

NOTES

1. Bruce Gagnon, interview by the author, January 2000.

2. U.S. Space Command, *Vision for 2020* (Peterson Air Force Base, Colorado, U.S. Space Command, 1996).

3. Ibid., p. 3.

4. Ibid., p. 4.

5. Ibid., p. 6.

6. Bill Sulzman, presentation at "Star Wars Revisited: An International Conference on Preventing an Arms Race in Space," organized by the Global Network against Weapons & Nuclear Power in Space, American University, Washington, D.C., April 15, 2000.

7. U.S. Space Command, *Long Range Plan*, (Peterson Air Force Base, Colorado, U.S. Space Command, 1998).

8. Ibid., p. 2.

9. Ibid., p. 20.

10. Ibid., p. vii.

11. Ibid., pp. vii–viii.

12. Ibid., p. viii.

13. Ibid., p. 1.

14. Ibid., pp. 1-3.

15. Ibid., p. 3.

16. Ibid., pp. 1-3.

17. U.S. Space Command, *Guardians of the High Frontier*, p. 4.

18. Ibid., p. 2.

19. U.S. Space Command, *Almanac 2000* (Peterson Air Force Base, Colorado), p. 32.

20. Ibid., p. 32.

21. William B. Scott, "USSC Prepares for Future Combat Missions in Space," *Aviation Week and Space Technology*, August 5, 1996, p. 51.

22. Ibid., p. 51.

23. Keith Hall, Speech to National Space Club, September 15, 1997.

24. "Implementing Our Vision for Space Control," General Richard B. Myers, speech to U.S. Space Foundation, Colorado Springs, April 7, 1999.

25. Team SBL News Release, "TRW-led Team SBL Awarded $10 Million Space Laser Contract," March 17, 1998.

26. Skip Vaughn, "Will We Land Space-Based Laser?" (Huntsville, Ala., Redstone Arsenal Public Affairs, 2000). Available at: http://www.redstone.army.mil/pub_affairs/000/03Mar2000/articles/0329100142806.html.

27. Quoted in "Megawatt Laser Test Brings Space Based Lasers One Step Closer," *Space Daily*, April 26, 2000. http://www.spacedaily.com/news/laser-00e.html.

28. Ibid.

29. Frances FitzGerald, *Way Out There in the Blue: Reagan, Star Wars and the End of the Cold War* (New York: Simon and Schuster, 2000), p. 499.

30. See United Nations Documentation: Research Guide, Resolutions adopted by the General Assembly at its fifty-fourth Session at: http://www.un.org/Depts/dhl/resguide/r54all1.htm. Resolution number is A/RES/54/53.

31. See United Nations Documentation: Research Guide, Resolutions Adopted by the General Assembly at its fifty-fifth Session at: http://www.un.org/Depts/dhl/resguide/r55all1.htm. Resolution number is A/RES/55/32.

32. Marc Vidricaire, "Outer Space," presentation at the First Committee of the General Assembly, October 1999.

33. Marc Vidricaire, "Presentation by Canada," October 19, 2000.

34. Vladimir V. Putin, "Address to the Millenium Summit," UN, September 6, 2000.

35. "Joint Statement of the Prime Minister of Canada and the President of the Russian Federation in the Sphere of Strategic Stability," Ottawa, Canada, November 20, 2000. Available at: http://www.dfait-maeci.gc.ca/english/geo/europe/russia-sphere-declaration-e.html.

36. Kofi Annan, speech before the UN Conference on Disarmament, Geneva, Switzerland, January 26, 1999, UN Press Release SG/SM/6874 DCF/354.

37. A record of the seminar was made by the Women's International League for Peace and Freedom: "Prevention of an Arms Race in Outer Space," *Report of the 1999 International Women's Day Seminar*, 10–11 March 1999, Palais des Nations, Geneva. WILPF's international office is at 1 rue de Varembe, C.P. 28, 1211 Geneva 20, Switzerland. E-mail: wilpf@iprolink.ch.

38. British media have been reporting on Menwith Hill. Journalist Duncan Campbell has authored many articles, among them "From Data Tapping to Missile Tracking: Star Wars Strikes Back," *Guardian*, December 2, 1998. Jonathon Carr-Brown's "United States Builds 'Son of Star Wars' at RAF Base" appeared in *Independent* on November 21, 1999. Both pieces and other articles on Menwith Hill are available through the Yorkshire Campaign for Nuclear Disarmament website at http://www.gn.apc.org/cndyorks/yspace/articles/.

39. Statement by Li Changhe, ambassador for disarmament affairs of China, in the Plenary Meeting of the Conference on Disarmament, March 11, 1999, Geneva.

40. Ibid.

41. Craig Eisendrath, interview by the author, May 1999.

42. Ibid.

43. Mike Moore, "Unintended Consequences," *The Bulletin of the Atomic Scientists*, January–February 2000, p. 64.

44. "Why the 21st Could Be the American Century," interview with George Friedman, *Parade Magazine*, April 6, 1997, p. 8.

45. George and Meredith Friedman, *Future of War: Power, Technology and American World Dominance in the 21st Century* (New York: Crown Publishers, 1996), p. 420.

46. U.S. Air Force Advisory Board, *New World Vistas: Air and Space Power for the 21st Century*, "Space Technology Volume," 1996, p. xviii.

47. Ibid., p. xviii.

48. Ibid., p. ix.

49. Ibid., p. 29.

50. Ibid., p. 29

51. Ibid., p. 29.

52. Ibid., p. 29.

53. Ibid., pp. 29–30.

54. Ibid., appendix B.

55. *Draft Environmental Impact Statement for Project Galileo*, (Pasadena, Calif. and NASA Headquarters, 1985), Jet Propulsion Laboratory, p. iii.

56. "First Part of Mapping Mission in Space Goes without a Hitch," *New York Times*, February 12, 2000.

57. Matthew Fordahl, Associated Press, "Red-Faced over the Red Planet: Metric Mix-up Doomed Mars Spacecraft," September 30, 1999.

58. Space Exploration, Power Sources for Deep Space Probes, U.S. General Accounting Office, May 1998, GAO/NSIAD-98-102, p. 3.

59. Interview, NASA Jet Propulsion Laboratory, Public Information Office, July 2000.

60. "ESA Unveils Its New Comet Chaser," European Space Agency "Information Note," 09–99, Paris, July 1, 1999, p. 2.

61. *Emergency Preparedness for Nuclear-Powered Satellites* (Paris: 1990), Organization for Economic Cooperation and Development and Swedish National Institute for Radiation Protection, p. 21.

62. Dr. John Gofman, interview by the author, January 1997.

63. *Emergency Preparedness for Nuclear-Powered Satellite*s, p. 24.

64. David L. Chandler, "Eyewitnesses in Chile Shed Light on Russian Probe's Spectacular Fall," *Boston Globe*, December 5, 1996.

65. *Final Environmental Impact Statement for the Cassini Mission*, National Aeronautics and Space Administration, Solar System Exploration Division, Office of Space Science, June 1995, p. 4–76.

66. Ibid., p. 4–72.

67. Dr. Ernest Sternglass, Interview in *Nukes in Space: The Nuclearization and Weaponization of the Heavens*, EnviroVideo, Box 311, Ft. Tilden, New York 11695, 1–800–ECO–TV46 (http://www.envirovideo.com).

68. "Memorandum of Understanding between the Department of Energy and the National Aeronautics and Space Administration Concerning Radioisotope Power Systems for Space Missions," signed July 26, 1991, by then NASA administrator Richard Truly and DOE secretary James Watkins.

69. "Space, The Cluttered Frontier," *Harper's Magazine*, December 1999.

70. National Research Council, *Protecting the Space Shuttle from Meteoroids and Orbital Debris*, (Washington, D.C.: National Academy Press, 1997), p. 4.

71. Ibid., p. 5.

72. "Space Junk's Out of Hand," *Newsday*, March 31, 1998.

73. Quoted in Nigel Flynn, *War In Space* (New York: Exeter Books, 1986), p. 6.

74. William J. Broad, *Teller's War: The Top-Secret Story behind the Star Wars Deception* (New York: Simon and Schuster, 1992), p. 105.

75. Flynn, *War in Space*, p. 8.

76. Ibid., p. 16.

77. William J. Broad, *Teller's War*, p. 108.

78. Flynn, p. 77.

79. Byron Spice, "SDI Looks to Nuclear Power, Orbiting Reactor Essential, Director Says," *Albuquerque Journal*, Jan. 12, l988.

80. Ibid.

81. Testimony of General James Abramson, "Space Nuclear Power, Conversion and Energy Storage for the Nineties and Beyond." Hearings before the Subcommittee on Energy Research and Production of the Committee on Science and Technology, United States House of Representatives, 99th Cong., 1st sess., October 8, 9,10, (United States Government Printing Office, Washington, D.C., 1985).

82. Christopher John Farley, "May the Shield Be with You: The Reagan-era Star Wars Program Lives on and May Yet Destabilize the New World Order," *Time*, July 10, 2000, p. 34.

83. Ibid.

84. "The Future of Space in the Military," speech by United States Senator Bob Smith to Aerospace Industry Association, Crystal City, Virginia, May 15, 2000.

85. Bruce Gagnon, interview by the author, June 2000.

86. John M. Collins, *Military Space Forces: The Next Fifty Years* (Washington, D.C.: Pergamon-Brasseys, 1989), p. 21.

87. Ibid., p. 23.

88. Ibid., p. 24.

89. Ibid., p. 27.

90. Ibid., p. 29.

91. Ibid., p. 32

92. Ibid., p. 32.

93. Ibid., p. 34.

94. Ibid., p. 36.

95. Ibid., pp. 42–43.

96. Ibid., p. 44.

97. Ibid., p. 48.

98. Ibid., p. 49.

99. Ibid., p. 54

100. Ibid., p. 57.

101. Ibid., back cover.

102. Ibid., back cover.

103. U.S. Senator Charles Robb, "Star Wars II," *Washington Quarterly*, winter 1999, pp. 84–85.

104. Cynthia McKinney, "Leave Start Wars to the Movies," *Congressional Record*, April 11, 2000, p. H2021.

105. Dennis Kucinich, "National Missile Defense," Congressional Record, April 11, 2000, p. H2020.

106. Dennis Kucinich, "Keynote Adress to the Global Network against Weapons and Nuclear Power in Space," April 15, 2000. A videotape of his address is available for $13 from the Global Network at PO Box 90083, Gainesville, FL 32607.

107. Ibid.

108. Mackubin Thomas Owens, "The Case for Missile Defense," *Wall Street Journal*, December 21, 2000.

109. Robert Burns, "Futuristic Weapons and Missile Defense to Fill Bush's Arsenal," *Star-Ledger*, September 24, 1999.

110. Walter Pincus, "Bush Nuclear Plans Could Face Hurdle; GOP-Backed Law Bars Proposed Steps,î *Washington Post*, June 4, 2000.

111. Mary McGrory, "Star Wars: Calling a Bomb a Bomb," *Washington Post*, July 13, 2000.

112. Center for Public Integrity, "Under the Influence, George W. Bush: Pragmatic, with Ties to Corporate America." Available at: wysiwg://32/htp://www.public-I-org/story_16_022899.htm.

113. William D. Hartung and Michelle Ciarrocca, "The Bush/Cheney Dream Team," *Multinational Monitor*, October 2000, p. 12.

114. Bruce Jackson, interview with author, December 2000.

115. Ibid.

116. Bill Mesler, "NATO's New Arms Bazaar: U.S Military Contractors and Diplomats Are Hawking Their Wares Together," *Nation*, July 21, 1997.

117. John Lancaster and Terry M. Neal, "Heavyweight 'Vulcans' Help Bush Forge a Foreign Policy," *Washington Post*, November 19, 1999.

118. Mesler, "NATO's Now Arms Bazaar."

119. Ibid.

120. Stephen J. Hadley, AFA National Convention Address, September 11, 2000, available at: http://www.afa.org/library/reports/hadley2k.html.

121. Walter Pincus, "From Missile Defense to a Space Arms Race," *Washington Post*, December 30, 2000.

122. "Text of Bush's Press Conference," Reuters, December 28, 2000.

123. For more information on the Center for Security Policy and Rumsfeld, see "Star Wars II, Here We Go Again" William Hartung and Michelle Ciarrocca, *Nation*, June 19, 2000 or, by the same authors, "The Tangled Web: The Marketing of Missile Defense 1994–2000," Arms Trade Resource Center Special Report, May 2000, available at: http://www.worldpolicy.org/projects/arms/reports/tangled.htm. For further details on Rumsfeld receiving the "Keeping of the Flame" award, see website of the Center for Security Policy: http://www/security-policy.org/flame1998.html.

124. See http://www.security-policy.org/board.html on the website of the Center for Security Policy.

125. Bruce Gagnon, interview by the author, December 2000.

126. Graham T. Allison, "ABC's of ABM and Missile Defense," *Christian Science Monitor*, May 31, 2000.

127. Noted on http://www/foreignrelations.org/public/resources/cgi?per!3351. A penetrating book on the council is Laurence H. Shoup and William Minter's *Imperial Brain Trust: The Council on Foreign Relations and United States Foreign Policy*, (New York: Monthly Review Press, 1977). "The Council on Foreign Relations is a key part of a network of people and institutions usually referred to by friendly observers as 'the establishment," they write. There are numerous websites listing the members of the Council on Foreign Relations.

128. See: http://www.geocities.com/Capitol, www.geocities.com/Capitol Hill/8425/TRI-CR.HTM or other websites that list members of the Council on Foreign Relations.

129. From the website for the Council on Foreign Relations: http://www.foreignrelations.org/public/about.html. The council's headquarters are at 58 East 68th St., New York, NY 10021.

130. Frank G. Klotz, Space, *Commerce, and National Security* (New York: Council on Foreign Relations, 1998), p. 56.

131. Jack Manno, *Arming the Heavens: The Hidden Military Agenda for Space*, 1945–1995 (New York: Dodd, Mead, 1984), p. 5.

132. Ibid., p. 11.

133. Ibid., p. 12.

134. Biography of Dr. Wernher Von Braun appears at: http://www.redstone.army.mil/history/vonbraun/bio.html.

135. Ibid., p. 13.

136. Ibid., p. 5.

137. Jack Manno, interview with author, July 2000.

138. Ibid.

139. Thomas L. Friedman, "What the World Needs Now: For Globalism to Work, America Can't Be Afraid to Act Like the Almighty Superpower That It Is," *New York Times Magazine*, March 28, 1999, p. 96.

140. Patricia M. Mische, *Star Wars and the State of Our Souls: Deciding the Future of Planet Earth* (Minneapolis: Winston Press,1985), p. v.

141. Global Network statement issued at press conference, July 20, 1992, Washington, D.C.

142. Regina Hagen, interview in *Nukes in Space 2: Unacceptable Risks*, EnviroVideo, Box 311, Ft. Tilden, New York 11965, 1–800–ECO–TV46, http://www.envirovideo.com.

143. Loring Wirbel, interview in *Nukes in Space 2: Unacceptable Risks.*

144. *Report of the Commission to Assess United States National Security Space Management and Organization*, (executive summary), p. 16, Washington, D.C., January 11, 2001.

145. Ibid., p. 10.

146. Ibid., p. 11.

147. Ibid., p. 12.

148. Ibid., p. A-4.

149. Ibid., p. 13.

150. *Report of the Commission to Assess United States National Security Space Management and Organization*, (full report), p. 33.

151. Ibid., p. 81.

152. Ibid., pp. 36–38, under the heading "C. Shape the International Legal and Regulatory Framework."

153. Staff Sgt. A. J. Bosker, "Air Force Welcomes Space Commission's Recommendations," Air Force Print News, February 2000, http://www.af.mil/news/Feb2001/n20010208_0185.shtml.

154. Jim Wolf, "Air Force Gearing Up for Space Operations," Reuters, February 9, 2001. Available at: http://dailynews.yahoo.com/h/nm/20010209/ ts/arms_space_dc_2.html.

155. Larry Wheeler, "Military Space Role Already Growing under Bush," *Florida Today*, February 9, 2001. Available at: http://www.space.com.com/missionlaunches/missions/fl_bush_010 209.html?Enews=y.

156. Kofi Annan, "Address on the Opening of the Third United Nations Conference on the Exploration and Peaceful Uses of Outer Space (UNISPACE III), July 19, 1999" (http://www.un.org/events/uni-space2/pressrel/e19am.htm).

157. Bruce Gagnon, interview with author, op. cit.

KARL GROSSMAN has specialized in investigative reporting for more than 35 years. Honors he has received include the George Polk, the James Aronson, and the John Peter Zenger Awards. His reporting on the use of nuclear power and plans to deploy weapons in space has been cited six times by Project Censored. Grossman is the author of many books, including *The Wrong Stuff: The Space Program's Nuclear Threat to Our Planet* and *Cover Up*: *What You Are Not Supposed to Know About Nuclear Power*. Grossman is a professor of journalism at the State University of New York, Old Westbury, where he teaches investigative reporting. He lives in tranquil Sag Harbor, New York.